Student Solutions Manual
Probability and Random Processes for Electrical Engineering
Second Edition

Student Solutions Manual
Probability and Random Processes for Electrical Engineering
Second Edition

Alberto Leon-Garcia
University of Toronto

▲▼ Addison-Wesley Publishing Company

Reading, Massachusetts
Menlo Park, California • New York
Don Mills, Ontario • Wokingham, England
Amsterdam • Bonn • Sydney • Singapore
Tokyo • Madrid • San Juan • Milan • Paris

Reproduced by Addison-Wesley from camera-ready copy supplied by the author.

Copyright © 1994 by Addison-Wesley Publishing Company, Inc.

All rights reserved. No part of this publication may be reproduced, stored in a retrieval system, or transmitted, in any form or by any means, electronic, mechanical, photocopying, recording, or otherwise, without the prior written permission of the publisher. Printed in the United States of America.

ISBN 0-201-55738-X

11 12 13 14 15 DA 07 06 05 04

Chapter 1

Probability Models in Electrical and Computer Engineering

1.1 a) In the first draw the outcome can be black (b) or white (w). If the first draw is black, then the second outcome can be b or w. However if the first draw is white, then the run only contains black balls so the second outcome must be b. Therefore $S = \{bb, bw, wb\}$.

b) In this case all outcomes can be b or w. Therefore $S = \{bb, bw, wb, ww\}$.

c) In part a) the outcome ww cannot occur so $f_{ww} = 0$. In part b) let N be a larger number of repetitions of the experiment. The number of times the first outcome is w is approximately $N/3$ since the run has one white ball and two black balls. Of these $N/3$ outcomes approximately $1/2$ are also white in the second draw. Thus $N/9$ if the outcome result is ww, and thus $f_{ww} = \frac{1}{9}$.

d) In the first experiment, the outcome of the first draw affects the probability of the outcomes in the second draw. In the second experiment, the outcome of the first draw does not affect the probability of the outcomes in the second draw.

1.4 The random experiment consisting of the toss of a coin has a sample space with two outcomes, $S = \{h, t\}$, and associated probability $p = P[\{h\}]$ and $1 - p = P[\{t\}]$.

a) In testing a device where the outcome is binary (i.e. "pass test" or "fail test") then the experiment corresponds to a coin toss where, say, "heads" corresponds to "pass test" and $P[h] = P[\text{pass test}] = $ long-term fraction of items that pass test.

b) The output of a fax machine is a series of black and white dots, so each output can be viewed as the outcome of a coin toss value, say, "black dot" corresponds to "heads" and $P[h] = $ [black dot] = long-term fraction of outputs that are black.

1.5 When the experiment is performed, either A occurs or it doesn't (i.e. B occurs); thus $N_A(n) + N_B(n) = n$ in n repetitions of the experiment, and

$$f_A(n) + f_B(n) = \frac{N_A(n)}{n} + \frac{N_B(n)}{n} = 1.$$

Thus $f_B(n) = 1 - f_A(n)$.

1.6 If A, B, or C occurs, then D occurs. Furthermore since A, B, or C cannot occur simultaneously, in n repetitions of the experiment we have

$$N_D(n) = N_A(n) + N_B(n) + N_C(n)$$

and dividing both sides by n

$$f_D(n) = f_A(n) + f_B(n) + f_C(n).$$

1.7
$$\begin{aligned}
<X>_n &= \frac{1}{n} \sum_{j=1}^{n} X(j) \quad n > 0 \\
&= \frac{n-1}{n} \frac{1}{n-1} \left\{ \sum_{j=1}^{n-1} X(j) + X(n) \right\} \\
&= \left(1 - \frac{1}{n}\right) <X>_{n-1} + \frac{1}{n} X(n) \\
&= <X>_{n-1} + \frac{X(n) - <X>_{n-1}}{n}
\end{aligned}$$

1.8 a) Assume that $X(j)$ assumes values from the sample space $S = \{x_1, x_2, \ldots, x,\}$, and let $N_k(n)$ be the number of tries x_k occurs in n repetitions of the experiment, then

$$< X^2 >_n = \frac{1}{n}\sum_{j=1}^{n} X^2(j)$$
$$= \frac{1}{N}\sum_{k=1}^{K} x_k^2 N_k(n)$$
$$\to \sum_{k=1}^{K} x_k^2 f_k(n)$$

Thus we expect that $< x^2 >_n \to \sum_{k=1}^{K} x_k^2 p_k$.

b) The same derivation of Problem 1.7, gives

$$< X^2 >_n = < X^2 >_{n-1} + \frac{X_n^2 - < X^2 >_{n-1}}{n}$$

1.10 a) Out of twenty repetitions of the experiment, the following outcomes led to "positive voltage": 7,3,4,7,4,3,4,5,4,1,3,1. The relative frequency of the event "positive voltage" is then $\frac{12}{20}$.

b)
$$< V >_{20} = \frac{1}{20}\{7+3-7+4+7-2-8+4+3+4-5+5+4+$$
$$1-6+3-7+1-9+0\}$$
$$= \frac{1}{20}\{2(-9)+1(-8)+1(-7)+1(-6)+1(-5)+1(-2)+$$
$$1(0)+2(1)+3(3)+4(4)+1(5)+2(7)\} = 0$$
$$< V^2 >_{20} = \frac{1}{20}\{2(-9)^2+1(-8)^2+...+2(7)^2\} = 27.8$$

1.13 Let T_s and T_p be the lifetime of the series and parallel systems in Fig. 1.11 respectively, and let T_i be the lifetime of the component C_i. The series system fails as soon as any of the n components fail, therefore T_s is the minimum of the lifetimes of the n outputs:

$$T_s = \min(T_1, ..., T_n) \ .$$

If the n components are in parallel configuration, the system fails when the last component fails, therefore

$$T_p = \max(T_1, ..., T_n) \ .$$

It then follows that:
$$T_s = \min(T_1, ..., T_n) \leq \max(T_1, ..., T_n) = T_p \quad .$$

Chapter 2

Basic Concepts of Probability Theory

2.1 Specifying Random Experiments

2.2 The outcome of this experiment consists of a pair of numbers (x,y) where x = number of dots in first toss and y = number of dots in second toss. Therefore, S = set of ordered pairs (x,y) where $x, y \epsilon \{1,2,3,4,5,6\}$ which are listed in the table below:

a)

x \ y	1	2	3	4	5	6
1	(1,1)	(1,2)	(1,3)	(1,4)	(1,5)	(1,6)
2	(2,1)	(2,2)	(2,3)	(2,4)	(2,5)	(2,6)
3	(3,1)	(3,2)	(3,3)	(3,4)	(3,5)	(3,6)
4	(4,1)	(4,2)	(4,3)	(4,4)	(4,5)	(4,6)
5	(5,1)	(5,2)	(5,3)	(5,4)	(5,5)	(5,6)
6	(6,1)	(6,2)	(6,3)	(6,4)	(6,5)	(6,6)

Checkmarks indicate elements of events below.

b)

x \ y	1	2	3	4	5	6
1	✓		✓		✓	
2		✓		✓		✓
3	✓		✓		✓	
4		✓		✓		✓
5	✓		✓		✓	
6		✓		✓		✓

A = "sum is even"

c)

x \ y	1	2	3	4	5	6
1						
2		✓		✓		✓
3						
4		✓		✓		✓
5						
6		✓		✓		✓

B = "both are even"

d) B is a subset of A so when B occurs then A also occurs, thus B implies A

e) $A \cap B^C = $ "sum is even and both tosses show odd number"

x	1	2	3	4	5	6
1	✓		✓		✓	
2						
3	✓		✓		✓	
4						
5	✓		✓		✓	
6						

f) $C = $ "number of dots differ by 1"

	1	2	3	4	5	6
1		✓				
2	✓		✓			
3		✓		✓		
4			✓		✓	
5				✓		✓
6					✓	

Comparing the tables for A and C we see that

$$A \cap C = \emptyset$$

2.5 a) Each testing of a pen has two possible outcomes: "pen good" (g) or "pen bad" b. The experiment consists of testing pens until a good pen is found. Therefore each outcome of the experiment consists of a string of "b's" ended by a "g". We assume that each pen is not put back in the drawer after being tested. Thus $\mathcal{S} = \{g, bg, bbg, bbbg\}$

b) We now simply record the number of pens tested, so $\mathcal{S} = \{1, 2, 3, 4\}$

c) The outcome now consists of a substring of "b's" and one "g" in any order, followed by a final "g". $\mathcal{S} = \{gg, bgg, gbg, gbbg, bbgg, gbbbg, bgbbg, bbgbg, bbbgg\}$

d) $\mathcal{S} = \{2, 3, 4, 5\}$

2.8 If we sketch the events A and B we see that $B = A \cup C$. We also see that the intervals corresponding to A and C have no points in common so $A \cap C = \emptyset$.

2.2. The Axioms of Probability

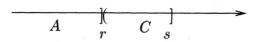

We also see that $(r, s] = (r, \infty) \cap (-\infty, s] = (-\infty, r]^c \cap (-\infty, s]$ that is $C = A^c \cap B$.

2.11

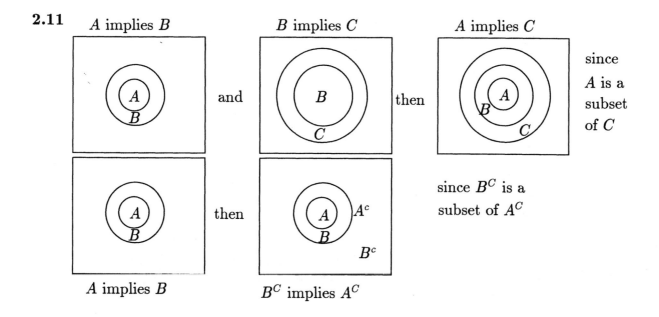

2.2 The Axioms of Probability

2.16 a) The sample space in tossing a die is $S = \{1, 2, 3, 4, 5, 6\}$. Let $p_i = P[\{i\}] = p$ since all faces are equally likely. By Axiom 1

$$\begin{aligned} 1 &= P[S] \\ &= P[\{1\} \cup \{2\} \cup \{3\} \cup \{4\} \cup \{5\} \cup \{6\}] \end{aligned}$$

The elementary events $\{i\}$ are mutually exclusive so by Corollary 4:

$$1 = p_1 + p_2 + \ldots + p_6 = 6p \Rightarrow p_i = p = \frac{1}{6} \text{ for } i = 1, \ldots, 6$$

b) We now have $p_1 = 2p$, $p_i = p$ for $i = 2, \ldots, 6$ so
$$1 = p_1 + \ldots + p_6 = 7p \Rightarrow p_1 = \frac{2}{7} \text{ and } p_i = p = \frac{1}{7} \text{ for } i = 2, \ldots, 6$$

c) $P[\text{even}] = P[\{2, 4, 6\}] = p_2 + p_4 + p_6$

$\therefore$ in Part a $P[\text{even}] = 3/6$
   in Part b $P[\text{even}] = 3/7$

2.21 Identities of this type are shown by application of the axioms. We begin by treating $(A \cup B)$ as a single event, then

$P[A \cup B \cup C]$
$= P[(A \cup B) \cup C]$
$= P[A \cup B] + P[C] - P[(A \cup B) \cap C]$ by Cor. 5
$= P[A] + P[B] - P[A \cap B] + P[C]$ by Cor. 5 on $A \cup B$
$\quad - P[(A \cap C) \cup (B \cap C)]$ and by distributive property
$= P[A] + P[B] + P[C] - P[A \cap B]$
$\quad - P[A \cap C] - P[B \cap C]$ by Cor. 5 on
$\quad + P[(A \cap B) \cap (B \cap C)]$ $(A \cap C) \cup (B \cap C)$
$= P[A] + P[B] + P[C] - P[A \cap B] - P[A \cap C]$ since
$\quad - P[B \cap C] + P[A \cap B \cap C].$ $(A \cap B) \cap (B \cap C) = A \cap B \cap C$

2.23 The sample space consists of 2^4 possible four-tuples with each component equal to "h" or "t". Following Example 27 we suppose that the 16 outcomes are equiprobable.

a) $P[A_2] = P[\{hhhh, thhh, hhth, hhht, thth, thht, hhtt, thtt\}]$
$= P[hhhh] + \ldots + P[thtt] = \frac{8}{16} = \frac{1}{2}$

b) $P[A_1 \cap A_3] = P[\text{tosses 1 and 3 are heads}]$
$= P[\{hhhh, hhht, hthh, htht\}] = \frac{4}{16} = \frac{1}{4}$

c) $P[A_1 \cap A_2 \cap A_3 \cap A_4] = P[\{hhhh\}] = \frac{1}{16}$

2.2. The Axioms of Probability

d) In the above we have been dealing with probabilities of intersections. This suggests using DeMorgans Rule:

$$\begin{aligned} P[A_1 \cup A_2 \cup A_3 \cup A_4] &= 1 - P[(A_1 \cup A_2 \cup A_3 \cup A_4)^c] \\ &= 1 - P[A^c \cap A_2^c \cap A_3^c \cap A_4^c] \\ &= 1 - P[\{tttt\}] \\ &= 1 - \frac{1}{16} = \frac{15}{16} \end{aligned}$$

2.27 Assume that the probability of any subinterval I of $[-1, 1]$ is proportional to its length, then
$$P[I] = k \text{ length } (I).$$
If we let $I = [-1, 1]$ then we must have that

$$1 = P[S] = P[[-1, 1]] = k \text{ length } ([-1, 1]) = 2k \Rightarrow k = \frac{1}{2}.$$

a) $P[A] = \frac{1}{2} \text{ length } ((-1, 0)) = \frac{1}{2}(1) = \frac{1}{2}$

$P[B] = \frac{1}{2} \text{ length } ((-0.5, 1)) = \frac{1}{2}\frac{3}{2} = \frac{3}{4}$

$P[C] = \frac{1}{2} \text{ length } ((0.75, 1)) = \frac{1}{2}\frac{1}{4} = \frac{1}{8}$

$P[A \cap B] = \frac{1}{2} \text{ length } ((-0.5, 0)) = \frac{1}{2}\frac{1}{2} = \frac{1}{4}$

$P[A \cap C] = P[\emptyset] = 0$

b) $P[A \cup B] \quad = P[S] = 1$

$P[A \cup C] \quad = \frac{1}{2} \text{ length } (A \cup C)$

$\quad = \frac{1}{2}\left(1 + \frac{1}{4}\right) = \frac{5}{8}$

$P[A \cup B \cup C] = P[S] = 1$

Now use axioms and corollaries:

$$P[A \cup B] = P[A] + P[B] - P[A \cap B] \quad \text{by Cor. 5}$$
$$= \frac{1}{2} + \frac{3}{4} - \frac{1}{4} = 1 \quad \checkmark$$
$$P[A \cap C] = P[A] + P[C] - P[\underbrace{A \cap C}_{\emptyset}] = \frac{1}{2} + \frac{1}{8} = \frac{5}{8} \quad \checkmark \quad \text{by Cor. 5}$$
$$P[A \cup B \cup C] = P[A] + P[B] + P[C]$$
$$- P[A \cap B] - P[A \cap C] - P[B \cap C]$$
$$+ P[A \cap B \cap C] \quad \text{by Eq. (2.7)}$$
$$= \frac{1}{2} + \frac{3}{4} + \frac{1}{8} - \frac{1}{4} - 0 - \frac{1}{8} + 0$$
$$= 1 \quad \checkmark$$

2.3 Computing Probabilities Using Counting Methods

2.34 Assume that on the first day the student can wear any of the four pairs; on any subsequent day he has a choice from 3 pairs, therefore

$$\text{number of distinct ordered 5-tuples} = 4 \cdot 3 \cdot 3 \cdot 3 \cdot 3 = 324$$

2.37 Imagine the sequence of students selecting desks. When there are 10 desks, the first student has 10 choices, the second student 9 choices and so on, thus

10 students and 10 desks: $10! = 3{,}628{,}800$

If there are 12 desks, the first student has 12 choices, the second has 11 choices, and thus:

10 students and 12 desks: $12(11)(10)(9)(8)(7)(6)(5)(4)(3) = 239{,}500{,}800$

2.4. Conditional Probability

2.41 The number of ways of choosing M out of 100 is $\binom{100}{M}$. This is the total number of equiprobable outcomes in the sample space.

We are interested in the outcomes in which m of the chosen items are defective and $M - m$ are nondefective.

The number of ways of choosing m defectives out of k is $\binom{k}{m}$.

The number of ways of choosing $M - m$ nondefectives out of $100 - k$ is $\binom{100 - k}{M - m}$.

The number of ways of choosing m defectives out of k <u>and</u> $M - m$ non-defectives out of $100 - k$ is
$$\binom{k}{m}\binom{100 - k}{M - m}$$

$$P[m \text{ defectives in } M \text{ samples}] = \frac{\text{\# outcomes with } k \text{ defective}}{\text{Total \# of outcomes}}$$

$$= \frac{\binom{k}{m}\binom{100 - k}{M - m}}{\binom{100}{M}}$$

This is called the <u>Hypergeometric</u> distribution.

2.4 Conditional Probability

2.48 a) The results follow directly from the definition of conditional probability:

$$P[A|B] = \frac{P[A \cap B]}{P[B]}$$

If $A \cap B = \emptyset$ then $P[A \cap B] = 0$ by Corollary 3 and thus $P[A|B] = 0$.

If $A \subset B$ then $A \cap B = A$ and $P[A|B] = \dfrac{P[A]}{P[B]}$.

If $A \supset B \Rightarrow A \cap B = B$ and $P[A|B] = \dfrac{P[B]}{P[B]} = 1$.

b) If $P[A|B] = \dfrac{P[A \cap B]}{P[B]} > P[A]$ then multiplying both sides by $P[B]$ we have:
$P[A \cap B] > P[A]P[B]$

We then also have that $P[B|A] = \dfrac{P[A \cap B]}{P[A]} > \dfrac{P[A]P[B]}{P[A]} = P[B]$.

We conclude that if $P[A|B] > P[A]$ then B and A tend to occur jointly.

2.52 The events B and C are shown below.

It then follows from the definition of conditional probability that:

$$P[B|C] = \dfrac{P[B \cap C]}{P[C]} = \dfrac{P[C]}{P[C]} = 1$$

$$P[C|B] = \dfrac{P[B \cap C]}{P[B]} = \dfrac{P[C]}{P[B]} = \dfrac{\frac{1}{4}}{\frac{3}{2}} = \dfrac{1}{6}$$

2.53 a) We use conditional probability to solve this problem. Let $A_i = \{$nondefective item found in ith test$\}$. A lot is accepted if the items in tests 1 and 2 are nondefective, that is, if $A_1 \cap A_2$ occurs. Therefore

$$P[\text{lot accepted}] = P[A_2 \cap A_1]$$
$$= P[A_2|A_1]P[A_1] \quad \text{by Eqn. 2.25}$$

This equation simply states that we must have A_1 occur, and then A_2 occur given that A_1 already occurred. If the lot of 100 items contains 5 defective items then

$$P[A_1] = \dfrac{95}{100} \quad \text{and}$$
$$P[A_2|A_1] = \dfrac{94}{99} \quad \text{since 94 of the many 99 items are defective.}$$

Thus

$$P[\text{lot accepted}] = \dfrac{94}{99} \cdot \dfrac{95}{100}.$$

2.4. Conditional Probability

If instead the lot contains 10 defective items we have
$$P[\text{lot accepted}] = P[A_2|A_1]P[A_1] = \frac{89}{99}\frac{90}{100}$$

b) If three items are tested and the lot is accepted when at most one of the tested items is defective, then the lot is accepted if $A_1 \cap A_2 \cap A_3$ or $A_1^c \cap A_2 \cap A_3$ or $A_1 \cap A_2^c \cap A_3$ or $A_1 \cap A_2 \cap A_3^c$ occurs. These four events are mutually exclusive (show it by taking the intersection of any pair) so by Corollary 4

$$\begin{aligned}P[\text{lot accepted}] &= P[A_1 \cap A_2 \cap A_3] + P[A_1^c \cap A_2 \cap A_3] \\ &+ P[A_1 \cap A_2^c \cap A_3] + P[A_1 \cap A_2 \cap A_3^c].\end{aligned}$$

Proceeding as in part a (but now dealing with a sequence of three events):
$$P[A_1 \cap A_2 \cap A_3] = P[A_3|A_1 \cap A_2]P[A_2|A_1]P[A_1]$$

Similarly we have
$$\begin{aligned}P[A_1^c \cap A_2 \cap A_3] &= P[A_3|A_1^c \cap A_2]P[A_2|A_1^c]P[A_1^c] \\ P[A_1 \cap A_2^c \cap A_3] &= P[A_3|A_1 \cap A_2^c]P[A_2^c|A_1]P[A_1] \\ P[A_1 \cap A_2 \cap A_3^c] &= P[A_3^c|A_1 \cap A_2]P[A_2|A_1]P[A_1]\end{aligned}$$

If 5 items in a lot of 100 are defective then
$$\begin{aligned}P[\text{lot accepted}] &= \frac{93}{98}\frac{94}{99}\frac{95}{100} + \frac{5}{100}\frac{95}{99}\frac{94}{98} + \frac{95}{100}\frac{5}{99}\frac{94}{98} + \frac{95}{100}\frac{94}{99}\frac{5}{98} \\ &= \frac{93}{98}\frac{94}{99}\frac{95}{100} + 3 \cdot \frac{5}{100}\frac{95}{99}\frac{94}{98}\end{aligned}$$

If 10 items in a lot are defective we have instead
$$P[\text{lot accepted}] = \frac{88}{98}\frac{89}{99}\frac{90}{100} + 3 \cdot \frac{10}{100}\frac{90}{99}\frac{89}{98}$$

2.59 a) We use the theorem on total probability to find $P[H]$:

$$\begin{aligned}P[H] &= P[H|\text{coin 1}]P[\text{coin 1}] + P[H|\text{coin 2}]P[\text{coin 2}] \\ &= p_1 \cdot \frac{1}{2} + p_2 \cdot \frac{1}{2} = \frac{1}{2}(p_1 + p_2)\end{aligned}$$

b) Using Bayes' Rule, we have
$$P[\text{coin 2}|H] = \frac{P[H|\text{coin 2}]P[\text{coin 2}]}{P[H]} = \frac{\frac{1}{2}p_2}{\frac{1}{2}(p_1 + p_2)} = \frac{p_2}{p_1 + p_2}$$

2.5 Independence of Events

2.62 The event A is the union of the mutually exclusive events $A \cap B$ and $A \cap B^c$, thus

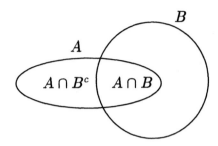

$$\begin{aligned}
P[A] &= P[A \cap B] + P[A \cap B^c] \quad \text{by Corollary 1} \\
\Longrightarrow P[A \cap B^c] &= P[A] - P[A \cap B] \\
&= P[A] - P[A]P[B] \quad \text{since } A \text{ and } B \text{ are independent} \\
&= P[A](1 - P[B]) \\
&= P[A]P[B^c] \Longrightarrow \quad A \text{ and } B^c \text{ are independent}
\end{aligned}$$

Similarly
$$P[B] = P[A \cap B] + P[A^c \cap B] = P[A]P[A] + P[A^c \cap B]$$

$$\Longrightarrow P[A^c \cap B] = P[B](1 - P[A]) = P[B]P[A^c]$$
$$\Rightarrow A \text{ and } B \text{ are independent}$$

Finally
$$P[A^c] = P[A^c \cap B] + P[A^c \cap B^c] = P[A^c]P[B] + P[A^c \cap B^c]$$

$$\Longrightarrow P[A^c \cap B^c] = P[A^c](1 - P[B]) = P[A^c]P[B^c]$$
$$\Rightarrow A^c \text{ and } B^c \text{ are independent}$$

2.65 We use a tree diagram to show the sequence of events. First we choose an urn, so A or A^c occurs. We then select a ball, so B or B^c occurs:

2.6. Sequential Experiments

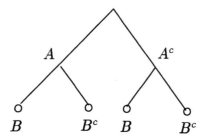

Now A and B are independent events if

$$P[B|A] = P[B]$$

But

$$P[B|A] = P[B] = P[B|A]P[A] + P[B|A^c]P[A^c]$$

$$\implies P[B|A](1 - P[A]) = P[B|A^c]P[A^c]$$
$$\implies P[B|A] = P[B|A^c] \quad \text{prob. of } B \text{ is the same given } A \text{ or } A^c, \text{ that is,}$$
the probability of B is the same for both urns.

2.6 Sequential Experiments

2.72 Let k be the number of defective items in a batch of n tested items, then k has binomial probabilities with parameters n and $p = \frac{1}{10}$. Using Corollary 1, we have:

$$\begin{aligned} P[k > 1] &= 1 - P[k \leq 1] = 1 - P[k = 0] - P[k = 1] \\ &= 1 - (1-p)^n - n(1-p)^{n-1}p, \quad p = \frac{1}{10} \\ &= 1 - \left(\frac{9}{10}\right)^n - n\left(\frac{9}{10}\right)^{n-1}\left(\frac{1}{10}\right) \end{aligned}$$

You should compare this problem to problem 53. How do the problems differ in the assumptions they make?

2.75 For an individual chip, the problem that the lifetime exceeds $\frac{1}{\alpha}$ seconds is:

$$P[\text{lifetime} > \tfrac{1}{\alpha}] = e^{-\alpha \frac{1}{\alpha}} = e^{-1} \triangleq p$$

If we consider the testing of whether each chip lifetime exceeds $\frac{1}{\alpha}$ as a sequence of Bernoulli trials, then

$$P[k \geq 5] = \sum_{k=5}^{10} \binom{10}{k} (e^{-1})^k (1 - e^{-1})^{10-k} = 0.289$$

2.79 a) A car pays k dollars if the time spent in the parking lot is in the range $\left(\frac{k-1}{2}, \frac{k}{2}\right]$, thus

$$\begin{aligned}
P[\text{pay } k \text{ dollars}] &= P\left[\frac{k-1}{2} < t \leq \frac{k}{2}\right] \\
&= P\left[t > \frac{k-1}{2}\right] - P\left[t > \frac{k}{2}\right] \quad \text{from Example 2.10} \\
&= e^{-\frac{k-2}{2}} - e^{-\frac{k}{2}} \\
&= \left(e^{\frac{1}{2}} - 1\right)\left(e^{-\frac{1}{2}}\right)^k \quad k = 1, 2, \ldots
\end{aligned}$$

Note this is a geometric probability law.

b) For $1 \leq k \leq 4$, $P[k]$ same as in part a), and for $k = 5$:

$$P[k = 5] = P\left[t > \frac{5-1}{2}\right] = e^{-2}$$

2.80 $P[k$ tosses required until heads comes up twice$] = P[$heads in kth toss$|1$ head in $k - 1$ tosses$]P[1$ head in $k - 1$ tosses$] = P[A|B]P[B]$.

Now $P[A|B] = P[1$ head in first $k - 1$ tosses$] = \binom{k-1}{1} p(1-p)^{k-2}$

Thus $P[A|B]P[B] = P[A|B]p = (k-1)p^2(1-p)^{k-2}$

2.84 Three tosses of a fair coin result in eight equiprobable outcomes:

$$\begin{array}{llll}
000 & \to 0 & 100 & \to 4 \\
001 & \to 1 & 101 & \to 5 \\
010 & \to 2 & \left.\begin{array}{l}101\\111\end{array}\right\} & \to \text{No output} \\
011 & \to 3 & &
\end{array}$$

2.6. Sequential Experiments

a) $P[\text{a number is output in step 1}] = 1 - P[\text{no output}]$
$$= 1 - \frac{2}{8} = \frac{3}{4}$$

b) Let $A_i = \{\text{output number } i\}$ $i = 0, ..., 5$
and $B = \{\text{a number is output in step 1}\}$
then

$$P[A_i|B] = \frac{P[A_i \cap B]}{P[B]} = \frac{P[\text{binary string corresponds to } i]}{\frac{3}{4}}$$
$$= \frac{\frac{1}{8}}{\frac{3}{4}} = \frac{1}{6}$$

c) Suppose we want to simulate an urn experiment with N equiprobable outcomes. Let n be the smallest integer such that $2^n \geq N$. We can simulate the urn experiment by tossing a fair coin n times and outputting a number when the binary string for the numbers $0, ..., N-1$ occur and not outputting a number otherwise.

2.87 a) The two outcomes are equiprobable.

b) If $U_n \leq (1-p)$, then $B_n = 0$; Otherwise $B_n = 1$.

c) Divide the unit interval into 6 disjoint subintervals of equal length and associate a face of the die with each subinterval.

d) For a finite number of outcomes, say n, divide the unit interval into n disjoint subintervals of length $p_1, p_2, ..., p_n$. This method is limited by the finite precision of the computer.

2.90 A Binomial random variable can be simulated by repeating the method in Problem 2.87b n times. A multinomial vector can be obtained by repeating the method in Problem 2.87d n times.

2.7 Problems Requiring Cumulative Knowledge

2.94 a)
$$P[k \text{ heads}] = P[k \text{ heads}|\text{coin 1}]P[\text{coin 1}] + P[k \text{ heads}|\text{coin 2}]P[\text{coin 2}]$$
$$= \binom{3}{k} p_1^k (1-p_1)^{3-k} \frac{1}{2} + \binom{3}{k} p_2^k (1-p_2)^{3-k} \frac{1}{2}$$

Using Bayes' Rule

$$P[\text{coin 1}|k \text{ heads}] = \frac{P[k|\text{coin 1}]P[\text{coin 1}]}{P[k]}$$

$$= \frac{\binom{3}{k} p_1^k (1-p_1)^{3-k} \frac{1}{2}}{\binom{3}{k} p_1^k (1-p_1)^{3-k} \frac{1}{2} + \binom{3}{k} p_2^k (1-p_2)^{3-k} \frac{1}{2}}$$

$$= \frac{p_1^k (1-p_1)^{3-k}}{p_1^k (1-p_1)^{3-k} + p_2^k (1-p_2)^{3-k}}$$

$P[\text{coin 2}|k \text{ heads}]$ is obtained by interchanging p_1 and p_2 in the above expression.

b)
$$P[\text{coin 1}|k] > P[\text{coin 2}|k]$$
$$\vdots \qquad \qquad \vdots$$
$$\frac{p_1^k (1-p_1)^{3-k}}{P[k]} > \frac{p_2^k (1-p_2)^{3-k}}{P[k]}$$
$$\iff p_1^k (1-p_1)^{3-k} > p_2^k (1-p_2)^{3-k}$$
$$\iff \left(\frac{p_1(1-p_2)}{p_2(1-p_1)}\right)^k > \left(\frac{1-p_2}{1-p_1}\right)^3 \quad \text{take log of both sides}$$
$$\iff k > \frac{3 \log \frac{1-p_2}{1-p_1}}{\log \frac{p_1(1-p_2)}{p_2(1-p_1)}} \quad \text{the last step assumes that } p_1 > p_2$$

c) For n tosses we have, coin 1 is more probable if

$$k > \frac{n \log \frac{1-p_2}{1-p_1}}{\log \frac{p_1(1-p_2)}{p_2(1-p_1)}}$$

Chapter 3

Random Variables

3.1 The Notion of a Random Variable

3.1 **a)** The sample space has 100 elements, with each element corresponding to a bill. $S = \{\xi_1, \xi_2, ..., \xi_{100}\}$ where ξ_i represents the ith bill. All bills are equiprobable:

$$P[\{\xi_i\}] = \frac{1}{100}$$

b) X is the denomination of a bill. There are three denominates so: $S_X = \{1, 5, 50\}$. The probability of a denomination proportional to the number of bills with that denomination:

$$P[X = 1] = P[\{\xi : X(\xi) = 1\}] = \frac{90}{100} = 0.90$$
$$P[X = 5] = P[\{\xi : (x\xi) = 5\}] = \frac{9}{100} = 0.09$$
$$P[X = 50] = P[\{\xi : X(\xi) = 50\}] = \frac{1}{100} = 0.01$$

3.2 The Cumulative Distribution Function

3.6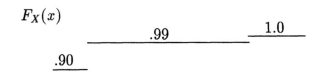

X is a discrete random variable.

3.14 a) X is a random variable of mixed type since it is continuous except for discontinuities at 0 and at 1.

b) $P\left[X < -\frac{1}{2}\right] = F_X\left(-\frac{1}{2}^-\right) = 0$

$P[X < 0] = F_X(0^-) = 0$ The point $x = 0$ is excluded from the event.

$P[X \leq 0] = F_X(0) = \frac{1}{4}$ The point $x = 0$ is included in the event.

$P\left[\frac{1}{4} \leq X < 1\right] = F_X(1^-) - F_X\left(\frac{1}{4}^-\right)$ Since $x = 1$ is excluded.

$= \frac{1}{2} - \left(\frac{1}{4} + \frac{1}{4}\left(\frac{1}{4}\right)\right) = \frac{3}{16}$ $x = \frac{1}{4}$ is included.

$P\left[\frac{1}{4} \leq X \leq 1\right] = F_X(1) - F_X\left(\frac{1}{4}^-\right) = 1 - \frac{5}{16} = \frac{11}{16}$

$P\left[X > \frac{1}{2}\right] = 1 - P\left[X \leq \frac{1}{2}\right] = 1 - F_X\left(\frac{1}{2}\right) = 1 - \left(\frac{1}{4} + \frac{1}{4}\frac{1}{2}\right) = \frac{5}{8}$

$P[X \geq 5] = 1 - P[X < 5] = 1 - F_X(5^-) = 0$

$P[X < 5] = F_X(5^-) = 1$

3.16 a) The cdf equals 1 at $\pi/2$, therefore $F_X\left(\frac{\pi}{2}\right) = 1 = c\left(1 + \sin\frac{\pi}{2}\right) = 2c \Rightarrow c = \frac{1}{2}$

3.3. The Probability Density Function

b)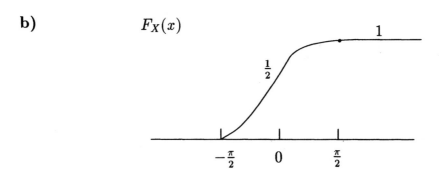

3.3 The Probability Density Function

3.19 a) We use the fact that the pdf must integrate to one:

$$1 = \int_0^1 f_X(x)dx = c\int_0^1 x(1-x)dx = c\left[\frac{x^2}{2} - \frac{x^3}{3}\right]_0^1 = \frac{c}{6}$$

$$\Rightarrow c = 6$$

b) $P\left[\frac{1}{2} \leq X \leq \frac{3}{4}\right] = 6\int_{1/2}^{3/4} x(1-x)dx = 6\left[\frac{x^2}{2} - \frac{x^3}{3}\right]_{1/2}^{3/4} = 0.34375$

c) For $x < 0$, $F_X(x) = 0$; for $x > 1$, $F_X(x) = 1$
For $0 \leq x \leq 1$

$$F_X(x) = \int_0^x f_X(x')dx' = 3x^2 - 2x^3$$

3.24 a) The pdf is obtained by differentiating $F_X(x)$. We obtain delta functions at points where $F_X(x)$ is discontinuous.

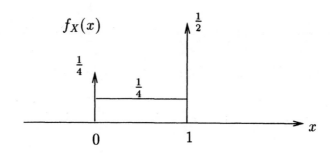

$$f_X(x) = \frac{1}{4}\delta(x) + \frac{1}{4}[u(x) - u(x-1)] + \frac{1}{2}\delta(x-1)$$

b) These probabilities are found by integrating the pdf. Care must be taken at the integral endpoints.

$$P[X < -\frac{1}{2}] = \int_{-\infty}^{-\frac{1}{2}} f_X(x)dx = 0$$

$$P[X < 0] = \int_{-\infty}^{0^-} f_X(x)dx = 0 \quad \text{note delta function at } x=0$$
$$\text{excluded from integral}$$

$$P[X \leq 0] = \int_{-\infty}^{0^+} f_X(x)dx = \frac{1}{4} \quad \text{Delta function at } x=0$$
$$\text{included in integral}$$

$$P[\frac{1}{4} \leq X \leq 1] = \int_{\frac{1}{4}}^{1^-} \frac{1}{4}dx + \int_{1^-}^{1^+} \frac{1}{2}\delta(x-1)dx$$
$$= \frac{3}{4}\frac{1}{4} + \frac{1}{2} = \frac{11}{16}$$

$$P[X > \frac{1}{2}] = \int_{\frac{1}{2}}^{\infty} f_X(x)dx = \int_{\frac{1}{2}}^{1^-} \frac{1}{4}dx + \int_{1^-}^{1^+} \frac{1}{2}\delta(x-1)dx$$
$$= \frac{1}{2}\frac{1}{4} + \frac{1}{2} = \frac{5}{8}$$

$$P[X \geq 5] = \int_{5^-}^{\infty} f_X(x)dx = 0$$

$$P[X < 5] = \int_{-\infty}^{5^-} f_X(x)dx = \int_{-\infty}^{0^-} 0\,dx + \int_{0^-}^{0^+} \frac{1}{4}\delta(x)dx$$
$$+ \int_{0^+}^{1^-} \frac{1}{4}dx + \int_{1^-}^{1^+} \frac{1}{2}\delta(x-1)dx + \int_{1^+}^{5^-} 0\,dx$$
$$= \frac{1}{4} + \frac{1}{4} + \frac{1}{2} = 1$$

3.28 a) From the definition of conditional probability we have:

$$F_X(x|a \leq X \leq b) = \frac{P[\{X \leq x\} \cap \{a \leq X \leq b\}]}{P[a \leq X \leq b]}$$

3.3. The Probability Density Function

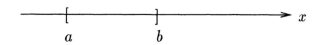

From the above figure we see that

$$\{X \le x\} \cap \{a \le X \le b\} = \begin{cases} \emptyset & \text{for } x < a \\ \{a \le X \le x\} & \text{for } a \le x \le b \\ \{a \le X \le b\} & \text{for } x > b \end{cases}$$

Therefore

$$F_X(x|a \le X \le b) = \begin{cases} \dfrac{P[\emptyset]}{P[a \le X \le b]} = 0 & x < a \\ \dfrac{P[a \le X \le x]}{P[a \le X \le b]} = \dfrac{F_X(x) - F_X(a^-)}{F_X(b) - F_X(a^-)} & a \le x \le b \\ \dfrac{P[a \le X \le b]}{P[a \le X \le b]} = 1 & x > b \end{cases}$$

b) $\quad f_X(x|a \le X \le b) = \dfrac{d}{dx} F_X(x|a \le X \le b)$

$$= \begin{cases} 0 & x < a \\ \dfrac{f_X(x)}{F_X(b) - F_X(a^-)} & a \le x \le b \\ 0 & x > b \end{cases}$$

Thus if X has pdf:

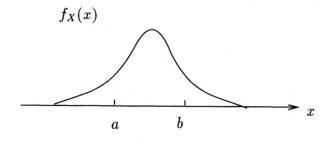

then $f_X(x|a \le X \le b)$ is

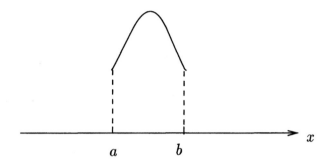

3.4 Some Important Random Variables

3.32 a) Let I_k denote the outcome of the kth Benoulli trial. The probability that the single event occurred in the kth trial is:

$$P[I_k = 1 | X = 1] = \frac{P[I_k = 1 \text{ and } I_j = 0 \text{ for all } j \neq k]}{P[X = 1]}$$

$$= \frac{P[0\ 0...\overset{k\text{th outcome}}{1}\ 0...0]}{P[X = 1]}$$

$$= \frac{p(1-p)^{n-1}}{\binom{n}{1} p(1-p)^{n-1}} = \frac{1}{n}$$

Thus the single event is equally likely to have occurred in any of the n trials.

b) The probability that the two successes occurred in trials j and k is:

$$P[I_j = 1, I_k = 1 | X = 2] = \frac{P[I_j = 1, I_k = 1, I_m = 0 \text{ for all } m \neq j, k]}{P[X = 2]}$$

$$= \frac{p^2(1-p)^{n-2}}{\binom{n}{2} p^2(1-p)^{n-2}} = \frac{1}{\binom{n}{2}}$$

3.4. Some Important Random Variables

Thus all $\binom{n}{2}$ possible choices of j and k are equally likely.

c) If $X = k$, then the location of successes is selected at random from among the $\binom{n}{k}$ possible permutations.

3.33 a) $\dfrac{p_k}{p_{k-1}} = \dfrac{\binom{n}{k} p^k q^{n-k}}{\binom{n}{k-1} p^{k-1} q^{n-k+1}} = \dfrac{\frac{n!}{k!(n-k)!}}{\frac{n!}{(k-1)!}} \cdot \dfrac{p}{q} = \dfrac{(n-k+1)p}{kq}$

$= \dfrac{(n+1)p - k(1-q)}{kq} = 1 + \dfrac{(n+1)p - k}{kq}$

b) First suppose $(n+1)p$ is not an integer, then for $0 \le k \le [(n+1)p] < (n+1)p$

$$(n+1)p - k > 0$$

so

$$\dfrac{p_k}{p_{k-1}} = 1 + \dfrac{(n+1)p - k}{kq} > 1$$

$\Rightarrow p_k$ increases as k increases from 0 to $[(n+1)p]$ for $k > (n+1)p \ge [(n+1)p]$

$$(n+1)p - k < 0$$

so

$$\dfrac{p_k}{p_{k-1}} = 1 + \dfrac{(n+1)p - k}{kq} < 1$$

$\Rightarrow p_k$ decreases as k increases beyond $[(n+1)p]$
$\therefore p_k$ attains its maximum at $k_{MAX} = [(n+1)p]$
If $(n+1)p = k_{MAX}$ then above implies that

$$\dfrac{p_{k_{MAX}}}{p_{k_{MAX}-1}} = 1 \Rightarrow p_{k_{MAX}} = p_{k_{MAX}-1}$$

3.38 $n = 10 \quad p = 0.1 \quad np = 1$

	$k = 0$	$k = 1$	$k = 2$	$k = 3$
Binomial	0.3487	0.387	0.1937	0.0574
Poisson	0.3679	0.3679	0.1839	0.0613

$n = 20 \quad p = 0.05 \quad np = 1$

	$k = 0$	$k = 1$	$k = 2$	$k = 3$
Binomial	0.3585	0.3774	0.1887	0.06
Poisson	0.3679	0.3679	0.1839	0.0613

$n = 100 \quad p = 0.01 \quad np = 1$

	$k = 0$	$k = 1$	$k = 2$	$k = 3$
Binomial	0.366	0.3697	0.1849	0.061
Poisson	0.3679	0.3679	0.1839	0.0613

We see that for $np =$ constant, as n increases and p decreases the accuracy of the approximation improves.

3.39 We use the Poisson approximation to binomial probabilities. $p = 0.001$ and $\lambda = np$. If $n = 50$ (i.e. no extra chips) $np = 0.05$ then

$$P[\text{all 50 chips working}] = \frac{\lambda^0}{k!}e^{-\lambda} = 0.951$$

so 50 chips do not give 99% probability of success. If $n = 51$, $np = 0.051$

$$P[\text{at least 50 chips working}] = \frac{\lambda^0}{k!}e^{-\lambda} + \frac{\lambda^1}{1!}e^{-\lambda}$$
$$= 0.9987$$

Thus 51 chips are enough to exceed 99% probability of success.

3.43
$$Q(-x) = \frac{1}{\sqrt{2\pi}}\int_{-x}^{\infty} e^{t^2/2}dt = 1 - \frac{1}{\sqrt{2\pi}}\int_{-\infty}^{-x} e^{-t^2/2}dt$$
$$= 1 - \frac{1}{\sqrt{2\pi}}\int_{\infty}^{x} e^{-t'^2/2}(-dt') \text{ where } t' = -t$$
$$= 1 - \frac{1}{\sqrt{2\pi}}\int_{x}^{\infty} e^{-t'^2/2}dt' = 1 - Q(x)$$

3.5 Functions of a Random Variable

3.52 No customers are left behind when $X \leq M$, therefore:

$$P[Y = 0] = P[(X - M)^+ = 0] = P[X \leq M]$$
$$= \sum_{k=1}^{M} p(1-p)^{k-1} = 1 - (1-p)^M$$

k customers are left behind when $X = M + k$, therefore:

$$P[Y = k] = P[X = M + k]$$
$$= p(1-p)^{M+k-1} \quad k = 1, 2, \ldots .$$

3.54 a) The equivalent event for $\{Y \leq y\}$ is $\{|X| \leq y\}$, therefore:

$$F_Y(y) = P[|X| \leq y] = P[-y \leq X \leq y]$$
$$= \begin{cases} 0 & y < 0 \\ F_X(y) - F_X(y^-) & y \geq 0 \end{cases}$$

Assuming X is a continuous random variable,

$$f_Y(y) = F_Y'(y) = f_X(y) + f_X(-y) \quad \text{for } y > 0 .$$

b) The equivalent event for $\{dy < Y \leq y + dy\}$ is shown below:

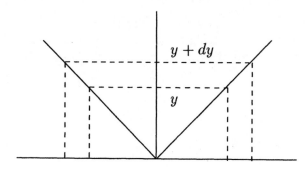

Therefore

$$P[y < Y \le y + dy] = P[y < X \le y + dy]$$
$$+ P[-y - dy < X \le -y]$$

$$\Rightarrow f_Y(y)dy = f_X(y)dy + f_X(-y)|dy|$$
$$\Rightarrow f_Y(y) = f_X(y) + f_X(-y) \quad \text{for } y > 0.$$

c) If $f_X(x)$ is an even function of x, then $f_X(x) = f_X(-x)$ and thus $f_Y(y) = 2f_X(y)$.

3.6 The Expected Value of Random Variables

3.63 $\mathcal{E}[X] = \dfrac{90}{100} \cdot 1 + \dfrac{9}{100} \cdot 5 + \dfrac{1}{100} \cdot 50 = 1.85$

The relative frequency interpretation implies that 1.85 is the long-term average payoff. Thus $E[X]$ is also the break-even price.

3.66 a) $\mathcal{E}[X] = \displaystyle\sum_{k=0}^{n} \binom{n}{k} p^k (1-p)^{n-k} k \quad$ Note $k=0$ term does not contribute to sum.

$$= \sum_{k=1}^{n} k \frac{n!}{k!(n-k)!} p^k (1-p)^{n-k}$$

$$= np \sum_{k=1}^{n} \frac{(n-1)!}{(k-1)!(n-k)!} p^{k-1} (1-p)^{n-k} \quad \text{let } k' = k-1$$

$$= np \underbrace{\sum_{k'=0}^{n-1} \frac{(n-1)!}{k'!(n-k')!} p^{k'} (1-p)^{n-1-k'}}_{\text{sum of all Binomial probs with parameter } n-1 \text{ and } p}$$

$$= np$$

$$\mathcal{E}[X^2] = \sum_{k=0}^{n} k^2 \frac{n!}{k!(n-k)!} p^k (1-p)^{n-k} \quad \text{cancelling } k \text{ and letting } k' = k-1$$

3.6. The Expected Value of Random Variables

$$= np \sum_{k'=0}^{n-1}(k'+1)\binom{n-1}{k'}p^{k'}(1-p)^{n-1-k}$$

$$= np\left\{\underbrace{\sum_{k'=0}^{n-1}k'\binom{n-1}{k'}p^{k'}(1-p)^{n-1-k}}_{\text{expected value of Binomial with parameters }n-1\text{ and }p} + \underbrace{\sum_{k'=0}^{n-1}\binom{n-1}{k'}p^{k'}(1-p)^{n-1-k'}}_{1}\right\}$$

$$= np\{(n-1)p+1\}$$

$$\sigma_X^2 = \mathcal{E}[X^2] - \mathcal{E}[X]^2 = np\{(n-1)p+1-np\} = np(1-p)$$

b) Since p is the long-term fraction of tosses that result in heads, it makes sense that the long-term fraction of heads in n tosses is np.

3.69
$$\mathcal{E}[X] = \int_{-\infty}^{\infty} x \frac{1}{\sqrt{2\pi}\sigma} e^{-(x-m)^2/2\sigma^2} = \frac{1}{\sqrt{2\pi}\sigma}\int_{-\infty}^{\infty}(y+m)e^{-y^2/2\sigma^2}dy$$

$$= \frac{1}{\sqrt{2\pi}\sigma}\int_{-\infty}^{\infty} y e^{-y^2/2\sigma^2}dy + \frac{m}{\sqrt{2\pi}\sigma}\int_{-\infty}^{\infty} e^{-y^2/2\sigma^2}dy$$

$$= \frac{-\sigma^2}{\sqrt{2\pi}\sigma}\left[e^{-y^2/2\sigma^2}\right]_{-\infty}^{\infty} + m = m$$

$$\sigma_X^2 = \int_{-\infty}^{\infty}(x-m)^2 \frac{e^{-(x-m)^2/2\sigma^2}}{\sqrt{2\pi}\sigma}dx = \frac{1}{\sqrt{2\pi}\sigma}\int_{-\infty}^{\infty} y^2 e^{-y^2/2\sigma^2}dy$$

$$= \frac{1}{\sqrt{2\pi}\sigma}\left\{\left[-\sigma^2 y e^{-y^2/2\sigma^2}\right]_{-\infty}^{\infty} + \sigma^2\int_{-\infty}^{\infty} e^{-y^2/2\sigma^2}dy\right\}$$

$$= \sigma^2 \quad \text{where we used integration by parts with}$$
$$u = y \quad dv = e^{-y^2/2\sigma^2}$$

3.77
$$\mathcal{E}[Y] = \int_{-\infty}^{\infty} g(x)f_X(x)dx \quad \text{write integral into three parts}$$

$$= -a\int_{-\infty}^{-a} f_X(x)dx + \int_{-a}^{a} x f_X(x)dx + a\int_{a}^{\infty} f_X(x)dx$$

$$= -aF_X(-a) + \int_{-a}^{a} x f_X(x)dx + a(1 - F_X(a^-))$$
$$\mathcal{E}[Y^2] = a^2 F_X(-a) + \int_{-a}^{a} x^2 f_X(x)dx + a^2(1 - F_X(a^-))$$
$$VAR[Y] = \mathcal{E}[Y^2] - \mathcal{E}[Y]^2$$

If X is Laplacian then $f_X(x) = \dfrac{\beta}{2} e^{-\beta |x|}$ and

$$\mathcal{E}[Y] = \int_{-\infty}^{\infty} \underbrace{g(x)}_{\text{odd fcn of } x} \underbrace{f_X(x)}_{\text{even fcn of } x} dx = 0$$

$$\begin{aligned}
VAR[Y] &= \mathcal{E}[Y^2] = a^2 e^{-\beta a} + \frac{1}{2} \int_{-a}^{a} x^2 \beta e^{-\beta x} dx \\
&= a^2 e^{-\beta a} + \beta \int_{0}^{a} x^2 e^{-\beta x} dx \\
&= a^2 e^{-\beta a} + \beta \left. \frac{e^{-\beta x}(\beta^2 x^2 + 2\beta x + 2)}{-\beta^3} \right|_0^a \quad \text{from integral tables in App. A} \\
&= \frac{2}{\beta^2} - \frac{2(1 + \beta a)}{\beta^2} e^{-\beta a}
\end{aligned}$$

3.7 The Markov and Chebyshev Inequalities

3.81 a) For a uniform random variable in $[-b, b]$ we have

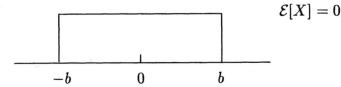

$\mathcal{E}[X] = 0$

Exact:
$$P[|X - m| > c] = \begin{cases} 1 - \frac{c}{b} & 0 \leq c \leq b \\ 0 & c > b \end{cases}$$

Chebyshev Bound gives
$$P[|X - m| > c] \leq \frac{\sigma_X^2}{c^2} = \frac{b^2}{3c^2}$$

3.7. The Markov and Chebyshev Inequalities

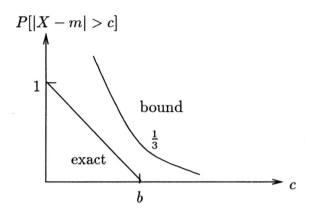

b) For the Laplacian random variable $\mathcal{E}[X] = 0$ and $VAR[X] = 2/\alpha^2$
Exact: $P[|X - m| > c] = P[|X| > c] = e^{-\alpha c}$
Bound: $P[|X - m| > c] \leq \dfrac{2}{\alpha^2 c^2}$

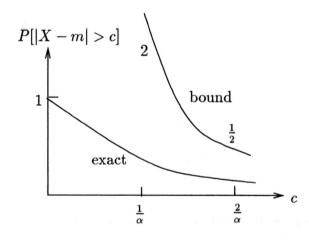

c) For the Gaussian random variable $\mathcal{E}[X] = 0$ and $VAR[X] = \sigma^2$
Exact: $P[|X - m| > c] = 2Q\left(\dfrac{c}{\sigma}\right)$
Bound: $P[|X - m| > c] \leq \dfrac{\sigma^2}{c^2}$

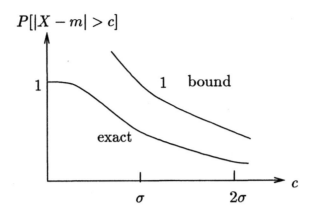

3.8 Testing the Fit of a Distribution to Data

3.84

	Obs.	Expected	$(0-\varepsilon)^2/\varepsilon$	
0	0	10.5	10.5	
1	0	10.5	10.5	
2	24	10.5	17.36	# degrees of freedom = 9
3	2	10.5	6.88	1% significance level $\Rightarrow$ 21.7
4	25	10.5	20.02	$D^2 > 21.7$
5	3	10.5	5.36	$\Rightarrow$ Reject hypothesis
6	32	10.5	44.02	that #'s are
7	15	10.5	1.93	unif. dist. in $\{0, 1, ..., 9\}$
8	2	10.5	6.88	
9	2	10.5	6.88	
	105		$D^2 = 130.33$	

	Obs.	Expected	$(0-\varepsilon)^2/\varepsilon$	
2	24	105/8	9.01	
3	2	105/8	9.43	# degrees of freedom = 9
4	25	105/8	10.74	1% significance level $\Rightarrow$ 21.7
5	3	105/8	7.81	$D^2 > 21.7$
6	32	105/8	77.41	$\Rightarrow$ Reject hypothesis
7	15	105/8	0.27	that #'s are
8	2	105/8	9.43	unif. dist. in $\{0, 1, ..., 9\}$
9	2	105/8	9.93	
	105		83.26	

3.9 Transform Methods

3.90
$$\begin{aligned}
E[X] &= \frac{1}{j}\frac{d}{d\omega}e^{jm\omega - \sigma^2\omega^2/2}\bigg|_{\omega=0} \\
&= \frac{1}{j}(jm - \sigma^2\omega)e^{jm\omega - \sigma^2\omega^2/2}\bigg|_{\omega=0} \\
&= m \\
E[X^2] &= \frac{1}{j^2}\frac{d^2}{d\omega^2}e^{jm\omega - \sigma^2\omega^2/2}\bigg|_{\omega=0} \\
&= \frac{1}{j^2}\left[-\sigma^2 e^{jm\omega - \sigma^2\omega^2/2} + (jm - \sigma^2\omega)^2 e^{jm\omega - \sigma^2\omega^2/2}\right]_{\omega=0} \\
&= \frac{1}{j^2}[-\sigma^2 + j^2 m^2] = \sigma^2 + m^2 \\
VAR[X] &= E[X^2] - \mathcal{E}[X]^2 = \sigma^2
\end{aligned}$$

3.93
$$\begin{aligned}
G_N(z) &= \sum_{k=0}^{n}\binom{n}{k}p^k(1-p)^{n-k}z^k \\
&= \sum_{k=0}^{n}\binom{n}{k}(pz)^k(1-p)^{n-k} \\
&= [pz + (1-p)]^n \quad \text{from Binomial Thm.}
\end{aligned}$$

3.98 $X = \begin{cases} X_1 & \text{with prob } p \\ X_2 & \text{with prob } 1-p \end{cases}$

$$\begin{aligned}
X^*(s) &= \mathcal{E}[e^{-sX}] = \mathcal{E}[e^{-sX}|X = X_1]p + \mathcal{E}[e^{-sX}|X = X_2](1-p) \\
&= \mathcal{E}[e^{-sX_1}]p + \mathcal{E}[e^{-sX_2}](1-p) \\
&= p\frac{\lambda_1}{s+\lambda_1} + (1-p)\frac{\lambda^2}{s+\lambda_2}
\end{aligned}$$

3.10 Basic Reliability Calculations

3.101 a) $f_T(t) = \begin{cases} \lambda e^{-\lambda(t-T_0)} & t \geq T_0 \\ 0 & t < T_0 \end{cases}$

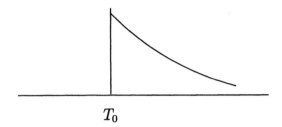

$$R(t) = P[T > t] = \begin{cases} 1 & 0 < t < T_0 \\ e^{-\lambda(t-T_0)} & t \geq T_0 \end{cases}$$

where we used the fact that

$$R(t) = P[T > t] = \int_t^\infty \lambda e^{-(\lambda t' - T_0)} dt' = e^{-\lambda(t-T_0)} \quad t > T_0$$

The $MTTF$ is given by the expected value of X:

$$\begin{aligned} MTTF = E[T] &= \int_0^\infty R(t') dt' \\ &= \int_0^{T_0} 1\, dt' + \int_{T_0}^\infty e^{-\lambda(t'-T_0)} dt' \\ &= T_0 + \frac{1}{\lambda} \end{aligned}$$

b) $r(t) = \dfrac{-R'(t)}{R(t)} = \begin{cases} 0 & 0 \leq t \leq T_0 \\ \lambda & t < T_0 \end{cases}$

c) $R(t) = e^{-\lambda(t-T_0)} = 0.99$

$$\Rightarrow \lambda(t - T_0) = \ln \frac{1}{0.99}$$

$$\Rightarrow t = T_0 + \frac{1}{\lambda} \ln \frac{1}{0.99} = T_0 + \frac{.01}{\lambda}$$

3.108 Each component has reliability: $R_i(t) = e^{-t}$

a) $R(t) = P[\text{system working at time}] = P[2 \text{ or more working at time } t]$
$$= \binom{3}{2}(e^{-t})^2(1-e^{-t}) + \binom{3}{3}(e^{-t})^3$$
$$= 3e^{-2t} - 2e^{-3t}$$

$$MTTF = \int_0^\infty R(t')dt' = \int_0^\infty (3e^{-2t'} - 2e^{-3t'})dt'$$
$$= \frac{3}{2} - \frac{2}{3} = \frac{5}{6}$$

b) Now $R_1(t) = R_2(t) = e^{-t}$ and $R_3(t) = e^{-t/2}$. $R(t) = P[2 \text{ or more working at time } t]$
$$= R_1(t)R_2(t)(1 - R_3(t)) + R_1(t)(1 - R_2(t))R_3(t)$$
$$+ (1 - R_1(t))R_2(t)R_3(t) + R_1(t)R_2(t)R_3(t)$$
$$= e^{-2t}(1 - e^{-t/2}) + 2e^{-t}(1 - e^{-t})e^{-t/2} + e^{-2t}e^{-t/2}$$
$$= e^{-2t} + 2e^{-3t/2} - 2e^{-5t/2}$$

$$MTTF = \int_0^\infty R(t')dt' = \int_0^\infty (3e^{-2t'} + 2e^{-\frac{3t}{2}} - 2e^{-\frac{5t'}{2}})dt'$$
$$= \frac{1}{2} + 2\frac{2}{3} - 2\frac{2}{5} = \frac{31}{30}$$

3.11 Generation of Random Variables

3.112
$$f_X(x) = \begin{cases} (a+x)/a^2 & -a \leq x \leq 0 \\ (a-x)/a^2 & 0 \leq x \leq a \\ 0 & \text{elsewhere} \end{cases}$$

$$\Rightarrow F_X(x) = \begin{cases} 0 & x < -a \\ \frac{1}{2}\left[\left(\frac{x}{a}\right)^2 + 2\left(\frac{x}{a}\right) + 1\right] & -a \leq x \leq 0 \\ \frac{1}{2} + \frac{1}{2}\left[2\left(\frac{x}{a}\right) - \left(\frac{x}{a}\right)^2\right] & 0 \leq x \leq a \\ 1 & x > a \end{cases}$$

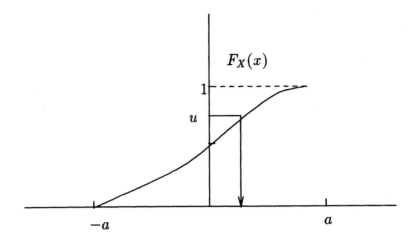

Solving the equation $u = F_X(x)$ for x we obtain

$$= F_X^{-1}(u) = \begin{cases} -a + a\sqrt{2u} & 0 \leq u \leq \frac{1}{2} \\ a - a\sqrt{2-2u} & \frac{1}{2} \leq u \leq 1 \end{cases}$$

3.115 If we search from 0 to 5, the average number of comparisons required is:

$$\begin{aligned} \mathcal{E}[N] &= 1P[0] + 2P[1] + ... + 6P[5] \\ &= \sum_{k=0}^{5}(k+1)\binom{5}{k}\left(\frac{1}{2}\right)^k\left(\frac{1}{2}\right)^{5-k} = 3.5 \end{aligned}$$

If we search in order of decreasing probability, we have

$$\begin{aligned} \mathcal{E}[N] &= 1P[3] + 2P[2] + 3P[4] + 4P[1] + 5P[5] + 6P[6] \\ &= 2.375 \end{aligned}$$

3.120 a) The key observation is that

$$P[X_1 \text{ is accepted}|X_1 = x] = \frac{f_X(x)}{Kf_W(x)}$$

since Y is uniform in $[0, Kf_W(x)]$. We then have that

3.12. Entropy

$$P[X_1 \text{ is accepted}] = \int_{-\infty}^{\infty} P[X_1 \text{ is accepted}|X_1 = x] f_W(x) dx$$
$$= \int_{-\infty}^{\infty} \frac{f_X(x)}{K f_W(x)} f_W(x) dx$$
$$= \frac{1}{K}$$

b) $P[x < X_1 < x + dx | X_1 \text{ accepted}]$
$$= \frac{P[\{X_1 \text{ accepted}\} \cap \{x < X_1 < x + dx\}]}{P[X_1 \text{ accepted}]}$$
$$= \frac{\frac{f_X(x)}{K f_W(x)} f_W(x) dx}{1/K}$$
$$= f_X(x) dx$$

$\therefore X_1$ when accepted as pdf $f_X(x)$ as desired.

3.12 Entropy

3.126 a) X takes on values from the set $\{1, 2, 3, 4, 5, 6\}$ with equal probabilities. Thus

$$H_X = -\sum_{k=1}^{6} \frac{1}{6} \log_2 \frac{1}{6} = \log_2 6$$

b) X now takes on values from $\{2, 4, 6\}$ with equal probabilities. Thus

$$H'_X = -\sum_{j=1}^{3} \frac{1}{3} \log_2 \frac{1}{3} = \log_2 3$$

The reduction in entropy is

$$H_X - H'_X = \log_2 6 - \log_2 3 = \log_2 \frac{6}{3} = 1 \text{ bit.}$$

3.131 The diagram below indicates what outputs are possible for each input.

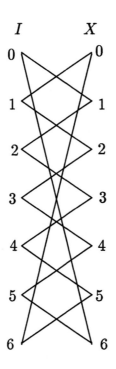

a) $H_I = -\sum_{j=0}^{6} \frac{1}{7} \log_2 \frac{1}{7} = \log_2 7.$

b) If $X = 4$ then $P[I = 3|X = 4] = P[I = 5|X = 9] = \frac{1}{2}.$ Therefore

$$H_{I|X=4} = -\frac{1}{2}\log_2 \frac{1}{2} - \frac{1}{2}\log_2 \frac{1}{2} = 1 \text{ bit}.$$

3.138 In Example 3.66 we divide the interval $[-a, a]$ into K subintervals of length Δ, that is $K\Delta = 2a$. The entropy of Q is:

$$\begin{aligned}
H_Q &= -\ln \Delta - \sum_{k=1}^{K} \frac{\Delta}{2a} \ln \frac{\Delta}{2a} \\
&= -\ln \Delta - \frac{K\Delta}{2a} \ln \frac{\Delta}{2a} \\
&= \ln 2a
\end{aligned}$$

On the other hand if $X > 0$ then $Q > 0$ and

3.12. Entropy

$$H_{Q/X>0} = -\ln\Delta - \sum_{k=1}^{K/2} \frac{\Delta}{a} \ln\frac{\Delta}{a}$$
$$= -\ln\Delta - \frac{K\Delta}{2a}\ln\frac{\Delta}{a}$$
$$= \ln a$$

The difference in entropy is

$$H_Q - H_{Q/X>0} = \ln 2a - \ln a = \ln 2 = 1 \text{ bit.}$$

From Eq. (3.111) the difference of differential entropies of X and $X > 0$ is:

$$\ln 2a - \ln a = 1 \text{ bit as well}$$

3.141 The entropy of X is:

$$\begin{aligned}H_X &= -\frac{3}{8}\log_2\frac{3}{8} - \frac{3}{8}\log_2\frac{3}{8} - \frac{1}{8}\log_2\frac{1}{8} - \frac{1}{16}\log_2\frac{1}{16} \\ &\quad -\frac{1}{32}\log_2\frac{1}{32} - \frac{1}{32}\log_2\frac{1}{32} \\ &= 1.999\end{aligned}$$

If we successively divide the set $\{1,2,3,4,5,6\}$ into subsets of approximately equal probability, we arrive at:

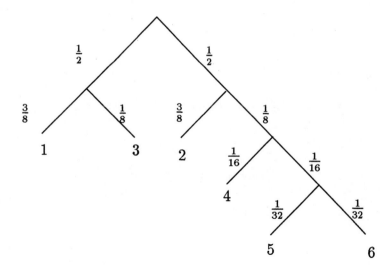

The average codeword length is

$$\mathcal{E}[L] = 2\left(\frac{3}{8}\right) + 2\left(\frac{1}{8}\right) + 2\left(\frac{3}{8}\right) + 3\left(\frac{1}{16}\right) + 4\left(\frac{2}{32}\right) = \frac{35}{16}$$

The following code is obtained by the "Huffman" Algorithm gives the least average codeword length:

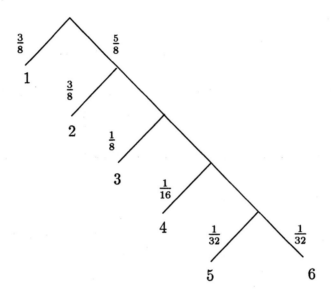

$$\mathcal{E}[L'] = \frac{3}{8} + 2\left(\frac{3}{8}\right) + 3\left(\frac{1}{8}\right) + 4\left(\frac{1}{16}\right) + 5\left(\frac{2}{32}\right) = \frac{33}{16}.$$

3.146 The maximum entropy random variable has pmf of form:

$$p_k = ce^{-\lambda k} \quad k = 1, 2, 3, 4$$

subject to the constraint

$$E[X] = \sum_{k=1}^{4} k p_k = 2.$$

Let $\alpha = e^{-\lambda}$ then the above equation becomes

$$2 = \sum_{k=1}^{4} ck\alpha^k = c(\alpha + 2\alpha^2 + 3\alpha^3 + 4\alpha^4)$$

3.12. Entropy

The constant c is determined from:

$$1 = \sum_{k=1}^{4} p_k = c(\alpha + \alpha^2 + \alpha^3 + \alpha^4)$$

Solving for c and substituting into the constraint equation we have

$$2 = \frac{\alpha + 2\alpha^2 + 3\alpha^3 + 4\alpha^4}{\alpha + \alpha^2 + \alpha^3 + \alpha^4} = \frac{1 + 2\alpha + 3\alpha^2 + 4\alpha^3}{1 + \alpha + \alpha^2 + \alpha^3}$$

Multiplying both sides by $1 + \alpha + \alpha^2 + \alpha^3$ and cancelling terms we arrive at an equation for α:

$$2(1 + \alpha + \alpha^2 + \alpha^3) = 1 + 2\alpha + 3\alpha^2 + 4\alpha^3$$
$$\Rightarrow 1 = \alpha^2 + 2\alpha^3$$

By trial and error we find $\alpha \approx 0.65$.

3.154 $Y = \max\{X_1, X_2, ..., X_n\}$.

$$\begin{aligned}
P[Y \leq y] &= P[X_1 \leq y, X_2 \leq y, ..., X_n \leq y] \quad 0 \leq y \leq a \\
&= P[X \leq y]^n \\
&= \left(\frac{y}{a}\right)^n \\
E[Y] &= \int_0^a y f_Y(y) dy = \int_0^1 y \frac{n y^{n-1}}{a^n} dy \\
&= \frac{n}{a^n} \left.\frac{y^{n+1}}{n+1}\right|_0^1 = \frac{n}{n+1} a \\
E[Y^2] &= \frac{n}{a^n} \int_0^1 y^2 y^{n-1} dy = \frac{n}{a^n} \left.\frac{y^{n+2}}{n+2}\right|_0^1 = \frac{n}{n+2} a^2 \\
VAR(Y) &= E[Y^2] - E[Y]^2 = \frac{n}{n+2} a^2 - \left(\frac{n}{n+1}\right)^2 a^2 \\
&= \left[\frac{n}{n+2} - \left(\frac{n}{n+1}\right)^2\right] a^2
\end{aligned}$$

The value "a" is by definition larger than any value Y can assume. In addition when n is larger the above results show that Y tends to be close to "a". Thus we conclude that Y is a good estimate for a.

Chapter 4

Multiple Random Variables

4.1 Vector Random Variables

4.1

a) $\{Y \geq X - 2\}$

b) $\{X < \ln b\}$

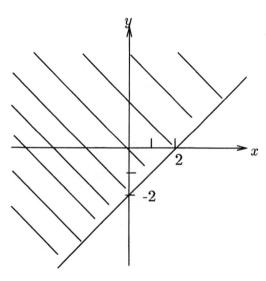

Not product-form

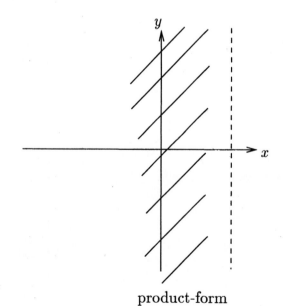

product-form

4.2 Pairs of Random Variables

4.4 a) All outcomes are equiprobable since: $P[X_1 = i, X_2 = j] = P[X_1 = i]P[X_2 = j] = \left(\frac{1}{6}\right)^2$, $1 \le j \le 6$, $1 \le j \le 6$

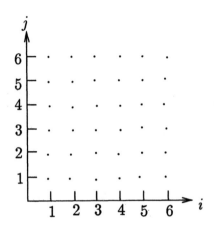

b) Since $X = \min(X_1, X_2)$ and $Y = \max(X_1, X_2)$ the joint proof of X and Y is:

$$P[X = i, Y = j] = \begin{cases} 0 & i > j \\ P[X_1 = i, X_2 = i] = \dfrac{1}{36} & i = j \\ P[X_1 = i, X_2 = j] + P[X_1 = j, X_2 = i] = \dfrac{2}{36} & i < j \end{cases}$$

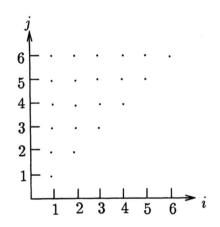

4.2. Pairs of Random Variables

c)
$$P[Y=6] = 5 \cdot \frac{2}{36} + \frac{1}{36} = \frac{11}{36}$$
$$P[Y=5] = 4 \cdot \frac{2}{36} + \frac{1}{36} = \frac{9}{36}$$
$$P[Y=4] = \frac{7}{36}$$
$$P[Y=3] = \frac{5}{36}$$
$$P[Y=2] = \frac{3}{36}$$
$$P[Y=1] = \frac{1}{36}$$

Similarly: $P[X=i] = P[Y=7-i] \qquad i=1,...,6$

4.8 a) For $0 \leq y_0 \leq x_0$ we integrate along the strip indicated below.

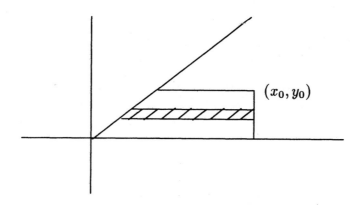

$$F_{XY}(x_0, y_0) = \int_0^{y_0} dy \int_y^{x_0} 2e^{-x}e^{-y} dx$$
$$= 1 - e^{-2y_0} - 2e^{-x_0}(1 - e^{-y_0})$$

For $0 < x_0 < y_0$ we now integrate as shown below

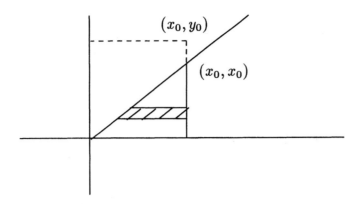

$$F_{XY}(x_0, y_0) = \int_0^{x_0} dy \int_y^{x_0} 2e^{-x}e^{-y}dx$$
$$= 1 - 2e^{-2x_0} - e^{-x_0}$$

b) The marginal cdf's are obtained by taking the appropriate limits of the joint cdf:

$$F_X(x_0) = \lim_{y_0 \to \infty} F_{XY}(x_0, y_0) = F_{XY}(x_0, x_0) = 1 - 2e^{-x_0} + e^{-2x_0}$$
$$F_Y(y_0) = \lim_{x_0 \to \infty} F_{XY}(x_0, y_0) = 1 - e^{-2y_0}$$

4.11 a)

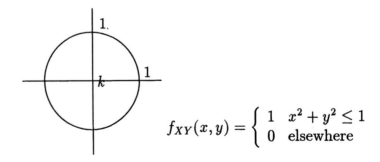

$$f_{XY}(x,y) = \begin{cases} 1 & x^2 + y^2 \leq 1 \\ 0 & \text{elsewhere} \end{cases}$$

a) To find k we note that

$$1 = \int_{-\infty}^{\infty}\int_{-\infty}^{\infty} f_{XY}(x,y)dxdy = \iint_{x^2+y^2 \leq 1} kdxdy = \pi$$

4.3. Independence of Two Random Variables

since area inside disk is $\pi(1)^2 = \pi \Rightarrow k = \frac{1}{\pi}$.

b) The marginal pdf for X is:

$$f_X(x) = \int_{-\sqrt{1-x^2}}^{\sqrt{1-x^2}} \frac{1}{\pi} dx = \frac{2\sqrt{1-x^2}}{\pi} \quad -1 \le x \le 1$$

Similarly, the marginal pdf for Y is:

$$f_X(y) = \frac{2\sqrt{1-y^2}}{\pi} \quad -1 \le y \le 1 \ .$$

4.12 a) The probability is obtained by integrating the joint pdf over the region indicated below.

$$\begin{aligned} P[X+Y \le 8] &= \int_0^8 \int_0^{8-x} 2e^{-x}e^{-2y} dy\, dx \\ &= \int_0^8 e^{-x}(1 - e^{-2(8-x)}) dx \\ &= 1 - 2e^{-8} + e^{-16} \end{aligned}$$

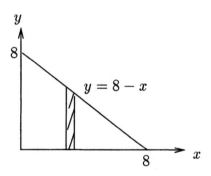

4.3 Independence of Two Random Variables

4.22 X and Y are independent random variables iff

$$f_{XY}(x,y) = f_X(x)f_Y(y) \quad \text{for all } x \text{ and } y$$

From 4.11 we have that for part (i)

$$f_X(x)f_Y(y) = \frac{4}{\pi^2}\sqrt{1-x^2}\sqrt{1-y^2} \quad \begin{array}{c} -1 \leq x \leq 1 \\ -1 \leq y \leq 1 \end{array}$$

Clearly this does not equal $f_{XY}(x,y)$ so X and Y are <u>not</u> independent.

4.24 a)
$$\begin{aligned} P[X^2 < \frac{1}{2}, |Y-1| < \frac{1}{2}] &= P[X^2 < \frac{1}{2}]P[\frac{1}{2} < Y < \frac{3}{2}] \\ &= \underbrace{P[X < \frac{1}{\sqrt{2}}]}_{\frac{1}{\sqrt{2}}} \underbrace{P[\frac{1}{2} < Y < \frac{3}{2}]}_{\frac{1}{2}} \\ &= \frac{1}{2\sqrt{2}} \ . \end{aligned}$$

4.4 Conditional Probability and Conditional Expectation

4.32 Again we'll do part (i) only

$$\begin{aligned} f_Y(y|x) &= \frac{f_{XY}(x,y)}{f_X(x)} = \frac{\frac{1}{\pi}}{\frac{2}{\pi}\sqrt{1-x^2}} \quad \text{for} \quad \begin{array}{c} -1\sqrt{1-x^2} < y < \sqrt{1-x^2} \\ -1 < x < 1 \end{array} \\ &= \frac{1}{2\sqrt{1-x^2}} \end{aligned}$$

Thus for a given x, Y is uniformly distributed in $[-\sqrt{1-x^2}, \sqrt{1-x^2}]$.

4.35 A customer is served by clerk C_i with probability

$$P[C_i] = p_i \ .$$

4.5. Multiple Random Variables

a) The cdf for the time T to service completion is

$$F_T(t) = P[T \leq t] = \sum_{i=1}^n P[T \leq t | C_i] P[C_i]$$
$$= \sum_{i=1}^n F_{T_i}(t) p_i$$

where T_i is an exponential random variable with parameter α_i.

The pdf is then

$$f_T(t) = \frac{d}{dt} F_T(t) = \sum_{i=1}^n f_{T_i}(t) p_i$$
$$= \sum_{i=1}^n \alpha_i e^{-\alpha_i t} p_i$$

b) We use conditional expectation to find the first two moments of T:

$$E[T] = E[E[T|C_i]]$$
$$= \sum_{i=1}^n E[T|C_i] p_i = \sum_{i=1}^n \frac{1}{\alpha_i} p_i$$
$$E[T^2] = E[E[T^2|C_i]] = \sum_{i=1}^n E[T_i^2|C_i] p_i$$
$$= \sum_{i=1}^n \frac{2}{\alpha_i^2} p_i$$

since the second moment of an exponential random variable is $2/\alpha_i^2$. Finally

$$VAR[T] = E[T^2] - E[T]^2$$
$$= \sum_{i=1}^n \frac{2 p_i}{\alpha_i^2} - \left(\sum_{i=1}^n \frac{p_i}{\alpha_i} \right)^2.$$

4.5 Multiple Random Variables

4.41 First find the conditional pdf of z given x and y:

$$f_{XYZ}(x,y,z) = f_Z(z|x,y)f_{XY}(x,y)$$
$$= f_Z(z|x,y)f_Y(y|x)f_X(x)$$

where in the last equality we used the conditional pdf for x, y.

4.43 a) Using the result from problem 41:

$$f_{X_1 X_2 X_3} = f_{X_1}(x_1) f_{X_2}(x_2|x_1) f_{X_3}(x_3|x_1, x_2)$$
$$= 1 \cdot \frac{1}{x_1} \cdot \frac{1}{x_2} \quad \text{for } 0 < x_3 < x_2 < x_1 < 1.$$

b) Here we need to carefully determine the limits of the integrals: For a given x_3, x_2 varies from x_3 to 1; for a given x_2, x_1 varies from x_2 to 1. Thus

$$f_{X_3}(x_3) = \int_{-\infty}^{\infty} \int_{-\infty}^{\infty} f_{X_1 X_2 X_3}(x_1, x_2, x_3) dx_1 dx_2$$
$$= \int_{x_3}^{1} dx_2 \int_{x_2}^{1} \frac{dx_1}{x_1 x_2} = \int_{x_3}^{1} \frac{dx_2}{x_2} \ln x_1 \big|_{x_2}^{1}$$
$$= \int_{x_3}^{1} (-1) \frac{\ln x_2}{x_2} dx_2 = (-1) \frac{(\ln x_2)^2}{2} \big|_{x_3}^{1}$$
$$= \frac{(-1)^2}{2} (\ln x_3)^2$$

$$f_{X_2}(x_2) = \int_{0}^{x_2} dx_3 \int_{x_2}^{1} \frac{dx_1}{x_1 x_2} = -\int_{0}^{x_2} dx_3 \frac{\ln x_2}{x_2} = -\ln x_2$$

We could also find the marginal pdf of X_2 by noting from the way the experiment is defined that:
$$f_{X_1 X_2}(x_1, x_2) = 1 \cdot \frac{1}{x_1} \quad 0 < x_2 < x_1 < 1$$

Thus
$$f_{X_2}(x_2) = \int_{x_2}^{1} f_{X_1 X_2}(x_1, x_2) dx_1 = \int_{x_2}^{1} \frac{dx_1}{x_1} = -\ln x_2 .$$

Clearly X_1 is uniform in $[0,1]$. Nevertheless we carry out the integral to find $f_{X_1}(x_1)$:

$$f_{X_1}(x_1) = \int_{0}^{x_1} dx_2 \int_{0}^{x_2} \frac{dx_3}{x_1 x_2}$$
$$= \int_{0}^{x_1} dx_2 \frac{1}{x_1} = 1 \quad 0 < x_1 < 1 .$$

4.6. Functions of Several Random Variables

c)
$$\begin{aligned}
f_{X_3}(x_3|x_1) &= \int_{x_3}^{x_1} f_{X_2 X_3}(x_2, x_3|x_1) dx_2 \\
&= \int_{x_3}^{x_1} \frac{f(x_1, x_2, x_3)}{f(x_1)} dx_2 \\
&= \int_{x_3}^{x_1} \frac{dx_2}{x_1 x_2} \\
&= \frac{1}{x_1}[\ln x_1 - \ln x_3] = \frac{1}{x_1} \ln \frac{x_1}{x_3} \qquad 0 < x_3 < x_1
\end{aligned}$$

d) An induction argument can be used to show that the integral of the joint pdf gives:
$$f_{X_n}(x_n) = \frac{(-1)^{n-1}(\ln x_n)^{n-1}}{(n-1)!} \qquad n \geq 1.$$

4.6 Functions of Several Random Variables

4.47 $T = \min(X_1, X_2, ..., X_n)$ where X_i Rayleigh. If the minimum is greater than t then all X_i's are greater than t:

$$\begin{aligned}
P[T > t] &= P[X_1 > t, X_2 > t, ..., X_n > t] \\
&= P[X_1 > t]P[X_2 > t]...P[X_n > t] \quad \text{by independence} \\
&= (e^{-t^2/2\alpha^2})^n \\
&= e^{-nt^2/2\alpha^2} \Rightarrow T \text{ is Rayleigh with parameter } \frac{\alpha}{\sqrt{n}} \\
\mathcal{E}[T] &= \sqrt{\frac{\pi}{2}} \frac{\alpha}{\sqrt{n}}
\end{aligned}$$

4.55 a) Let the auxiliary function be $W = Y$ then $X = ZW/(1-Z)$ and

$$J(z,w) = \begin{vmatrix} \frac{w}{(1-z)^2} & \frac{z}{1-z} \\ 0 & 1 \end{vmatrix} = \frac{w}{(1-z)^2}$$

and

$$f_{Z,W}(z,w) = f_{XY}\left(\frac{zw}{1-z}, w\right) \frac{|w|}{(1-z)^2}$$

so

$$f_Z(z) = \int_{-\infty}^{\infty} f_{XY}\left(\frac{zw}{1-z}, w\right) \frac{|w|}{(1-z)^2} dw$$

b) $\quad f_Z(z) = \int_0^{\infty} \alpha e^{-\alpha zw/(1-z)} \alpha e^{-\alpha w} \frac{w}{(1-z)^2} dw \quad 0 \le z \le 1$

$\quad\quad\quad = \frac{\alpha^2}{(1-z)^2} \int_0^{\infty} w e^{-\frac{\alpha}{(1-z)}w} = \frac{\alpha^2}{(1-z)^2} \frac{(1-z)^2}{\alpha^2}$

$\quad\quad\quad = 1$

Thus Z is uniformly distributed in $[0,1]$.

4.7 Expected Value of Functions of Random Variables

4.59 a) $\mathcal{E}[(X+Y)^2] = \mathcal{E}[X^2 + 2XY + Y^2] = \mathcal{E}[X^2] + 2\mathcal{E}[XY] + \mathcal{E}[Y^2]$

b) $VAR[X+Y] = \mathcal{E}[(X+Y)^2] - \mathcal{E}[X+Y]^2$
$\quad\quad\quad\quad\quad\quad = \mathcal{E}[X^2] + 2\mathcal{E}[XY] + \mathcal{E}[Y^2] - \mathcal{E}[X]^2 - 2\mathcal{E}[X]\mathcal{E}[Y] - \mathcal{E}[Y]^2$
$\quad\quad\quad\quad\quad\quad = VAR[X] + VAR[Y] + 2[\mathcal{E}[XY] - \mathcal{E}[X]\mathcal{E}[Y]]$

c) $VAR[X+Y] = VAR[X] + VAR[Y]$ if $\mathcal{E}[XY] = \mathcal{E}[X]\mathcal{E}[Y]$
that is, if X and Y are <u>uncorrelated</u>.

4.64 a) i) $\mathcal{E}[XY] = (-1)^2 \frac{1}{6} + 2(1)(-1)\frac{1}{6} + 1^2 \frac{1}{6} = 0 \Rightarrow \rho_{XY} = 0$

b) i) $\hat{Y} = 0$

c) i) $p(y|-1) = \frac{1}{2}$ for $y = -1, 1 \Rightarrow \hat{Y} = 1$ or $\hat{Y} = -1$
$\quad\quad p(y|0) = 1 \quad\quad$ for $y = 0 \quad\quad\quad \Rightarrow \hat{Y} = 0$
$\quad\quad p(y|1) = \frac{1}{2} \quad$ for $y = -1, 1 \Rightarrow \hat{Y} = 1$ or $\hat{Y} = -1$

4.66 i) From the solution to Prob. 4.11 we have:

4.7. Expected Value of Functions of Random Variables

$$\mathcal{E}[X] = \mathcal{E}[Y] = 0$$
$$\mathcal{E}[XY] = \int_{-1}^{1} \int_{-\sqrt{1-x^2}}^{\sqrt{1-x^2}} x \frac{1}{\pi} dy dx = 0$$
$$\Rightarrow \rho = 0$$

4.71 a) From Ex. 4.36 we have that V and W are independent zero-mean Gaussian random variables with respective variances $1+\rho$ and $1-\rho$. Thus their joint characteristic function is

$$\begin{aligned}
\Phi_{VW}(w_1, w_2) &= E[e^{j(w_1 V + w_2 W)}] \\
&= E[e^{jw_1 V}] E[e^{jw_2 W}] \\
&\quad \text{since } V \text{ and } W \text{ are independent and Ex. 4.40} \\
&= \Phi_V(w_1) \Phi_W(w_2) \\
&= e^{j(1+\rho)w_1^2} e^{j(1-\rho)w_2^2} \\
&\quad \text{from the characteristic function of a} \\
&\quad \text{Gaussian random variable.}
\end{aligned}$$

Since $X = (V - W)/\sqrt{2}$ and $Y = (V + W)/\sqrt{2}$, we have

$$\begin{aligned}
\Phi_{XY}(u_1, u_2) &= E[e^{j(u_1 X + u_2 Y)}] \quad \text{substitute for } X \text{ and } Y \\
&= E\left[e^{j\left(u_1 \frac{V-W}{\sqrt{2}} + u_2 \frac{V+W}{\sqrt{2}}\right)}\right] \\
&\quad \text{rearrange in terms of } V \text{ and } W \\
&= E\left[e^{j\left[\left(\frac{u_1}{\sqrt{2}} + \frac{u_2}{\sqrt{2}}\right)V + \left(-\frac{u_1}{\sqrt{2}} + \frac{u_2}{\sqrt{2}}\right)W\right]}\right] \\
&\quad \text{this is the joint characteristic function of } V \text{ and } W \\
&= \Phi_{VW}\left(\frac{u_1}{\sqrt{2}} + \frac{u_2}{\sqrt{2}}, -\frac{u_1}{\sqrt{2}} + \frac{u_2}{\sqrt{2}}\right) \quad \text{from above} \\
&= e^{j(1+\rho)(u_1+u_2)^2/2} e^{j(1-\rho)(u_2-u_1)^2/2}
\end{aligned}$$

b) $\Phi_{XY}(u_1, u_2) = e^{j[u_1^2 + u_2^2 - 2\rho u_1 u_2]} \triangleq e^{\{\cdot\}}$

$$\frac{\partial \Phi_{XY}}{\partial u_1} = e^{\{\cdot\}} j[2u_1 - 2\rho u_2]$$

$$\frac{\partial^2 \Phi_{XY}}{\partial u_1^2} = e^{\{\cdot\}} j^2 [2u_1 - 2\rho u_2]^2 + e^{\{\cdot\}} j2$$

$$\frac{\partial \partial^2 \Phi_{XY}}{\partial u_2 \partial u_1^2} = e^{\{\cdot\}}j^3[2u_1 - 2\rho u_2]^2[2u_2 - 2\rho u_1]$$
$$+ e^{\{\cdot\}}j^2 2[2u_1 - 2\rho u_2][-2\rho]$$
$$+ e^{\{\cdot\}}j[2u_2 - 2\rho u_1]j2$$
$$= 0 \quad \text{when evaluated at } u_1 = u_2 = 0.$$

Thus $E[X^2 Y] = 0$.

c)
$$\Phi_{X'Y'}(u_1, u_2) = E[e^{j(u_1 X' + u_2 Y')}]$$
$$= E[e^{j(u_1 X + u_1 a + u_2 Y + u_2 b)}]$$
$$= e^{j(u_1 a + u_2 b)} E[e^{j(u_1 X + u_2 Y)}]$$
$$= e^{j(u_1 a + u_2 b)} \Phi_{XY}(u_1, u_2).$$

4.8 Jointly Gaussian Random Variables

4.76 The solution involves matching the coefficients of the polynomials in the exponent of the Gaussian pdf. If the exponent is given as:

$$-\{ax^2 + by^2 + cxy + dx + ey + f\}$$

then from Eqn. 4.79 we must have

1. coeff of $x^2 \Rightarrow \dfrac{1}{2(1-\rho^2)\sigma_1^2} = a \Rightarrow 1' \quad \sigma_1^2 = \dfrac{1}{2(1-\rho^2)a}$

2. coeff of $y^2 \Rightarrow \dfrac{1}{2(1-\rho^2)\sigma_2^2} = b \Rightarrow 2' \quad \sigma_2^2 = \dfrac{1}{2(1-\rho^2)b}$

3. coeff of $xy \Rightarrow \dfrac{-2\rho}{2(1-\rho^2)\sigma_1 \sigma_2} = c$

1 & 2 $\Rightarrow \dfrac{\sigma_2^2}{\sigma_1^2} = \dfrac{a}{b}$

1,2 & 3 $\Rightarrow \dfrac{-2\rho a}{\sqrt{\frac{a}{b}}} = c \Rightarrow \rho = -\dfrac{c}{2\sqrt{ab}}$

4.8. Jointly Gaussian Random Variables

To find m_1 and m_2 we solve the following two equations in m_1 and m_2:

$$d = -\frac{2m_1}{2(1-\rho^2)\sigma_1^2} + \frac{2\rho m_2}{2(1-\rho^2)\sigma_1\sigma_2} = -2am_1 - cm_2$$

$$e = -\frac{2m_2}{2(1-\rho^2)\sigma_2^2} + \frac{2\rho m_1}{2(1-\rho^2)\sigma_1\sigma_2} = -2bm_2 - cm_1$$

Not all choices of a, b, c, d, e and f lead to legitimate Gaussian pdf's. An unfortunate error in the first printing resulted in an invalid choice. We have

$$a = \frac{1}{2} \quad b = 1 \quad c = 2$$

then

$$\rho = -\frac{2}{2\sqrt{1/2}} = -\sqrt{2}$$

But a valid ρ must satisfy $|\rho| < 1$, so we see the above choice is invalid.

If instead we had been given the exponent

$$-\{\frac{2}{3}x^2 + \frac{8}{3}y^2 + \frac{4}{3}xy - 4x - 8y + 8\}$$

we would have:

$$\rho = -\frac{c}{2\sqrt{ab}} = -\frac{\frac{4}{3}}{2\sqrt{\frac{2}{3} \cdot \frac{8}{3}}} = -\frac{1}{2}$$

$$\sigma_1^2 = \frac{1}{2(1-\rho^2)a} = 1 \qquad \sigma_2^2 = \frac{1}{2(1-\rho^2)b} = \frac{1}{4}$$

and

$$\left.\begin{array}{r}-2\left(\frac{2}{3}\right)m_1 - \frac{4}{3}m_2 = -4 \\ -2\left(\frac{8}{3}\right)m_2 - \frac{4}{3}m_1 = -8\end{array}\right\} \Rightarrow \begin{array}{l}m_1 = 2 \\ m_2 = 1\end{array}$$

4.82 First we find the conditional second moment of X:

$$\mathcal{E}[X^2|y] = VAR[X|y] + \mathcal{E}[X|y]^2 \quad \text{where we assume } \mathcal{E}[X] = \mathcal{E}[Y] = 0$$

$$= \sigma_X^2(1-\rho^2) + \left(\rho\frac{\sigma_X}{\sigma_Y}y\right)^2$$

where we used the conditional pdf of X. Now we use conditional expectation:

$$\begin{aligned}
\mathcal{E}[X^2Y^2] &= \mathcal{E}[\mathcal{E}[X^2Y^2|Y]] = \mathcal{E}[Y^2\mathcal{E}[X^2|Y]] \\
&= \mathcal{E}[\sigma_X^2(1-\rho^2)Y^2 + \rho^2\frac{\sigma_X^2}{\sigma_Y^2}Y^4]
\end{aligned}$$

where we used $E[X^2|Y]$ from above

$$= \sigma_X^2\sigma_Y^2(1-\rho^2) + \rho^2\frac{\sigma_X^2}{\sigma_Y^2}\underbrace{\mathcal{E}[Y^4]}_{3\sigma_Y^2}\} \text{ shown below}$$

$$\begin{aligned}
&= \sigma_X^2\sigma_Y^2(1+2\rho^2) \\
&= \sigma_X^2\sigma_Y^2 + 2\mathcal{E}[XY] = \mathcal{E}[X^2]\mathcal{E}[Y^2] + 2\mathcal{E}[XY]
\end{aligned}$$

$$\begin{aligned}
\mathcal{E}[Y^4] &= \int_{-\infty}^{\infty} \frac{y^4 e^{-y^2/2\sigma^2}}{\sqrt{2\pi}\sigma}dy \quad t = \frac{y}{\sigma} \\
&= \frac{\sigma^5}{\sqrt{2\pi}\sigma}\underbrace{2\int_0^{\infty} t^4 e^{-t^2/2}dt}_{\frac{\Gamma(\frac{5}{2})}{2\alpha^5}} \quad \begin{array}{l}\text{from Table in Appendix A}\\ \text{where } \alpha^2 = \frac{1}{2}\end{array}
\end{aligned}$$

But $\Gamma\left(\frac{5}{2}\right) = \frac{3}{2}\Gamma\left(\frac{3}{2}\right) = \frac{3}{2}\frac{1}{2}\Gamma\left(\frac{1}{2}\right) = \frac{3}{4}\sqrt{\pi}$, thus

$$\begin{aligned}
E[Y^4] &= \frac{\sigma^5}{\sqrt{2\pi}\sigma}\frac{2\frac{3}{4}\sqrt{\pi}}{2\left(\frac{1}{\sqrt{2}}\right)^5} \\
&= 3\sigma^4
\end{aligned}$$

4.88 The joint characteristic function for (X_1, X_2, X_3, X_4) is:

$$\Phi_{\underline{X}}(\underline{w}) = e^{-\frac{1}{2}\underline{w}^T K \underline{w}}$$

where

$$\underline{w}^T K \underline{w} = (w_1, w_2, w_3, w_4)\left[E[X_iX_j]\right]\begin{bmatrix}w_1\\w_2\\w_3\\w_4\end{bmatrix} = \sum_{i=1}^{4}\sum_{j=1}^{4}E[X_iX_j]w_iw_j$$

Expanding the exponential in a power series:

$$\Phi_{\underline{X}}(\underline{w}) = 1 - \frac{1}{2}\underline{w}^T K \underline{w} + \frac{1}{8}(\underline{w}^T K \underline{w})^2 + \ldots$$

From the moment theorem we know that $E[X_1X_2X_3X_4]$ is the coefficient of $w_1w_2w_3w_4$ in the above series. This coefficient appears in the third term:

$$\frac{1}{8}(\underline{w}^T K \underline{w})^2 = \frac{1}{8}\left(\sum_{ij} E[X_iX_j]x_ix_j\right)\left(\sum_{i'j'} E[X_{i'}X_{j'}]w_{i'}w_{j'}\right)$$

$$= \frac{1}{8}\sum_{ij}\sum_{i'j'} E[X_iX_j]E[X_{i'}X_{j'}]w_iw_jw_{i'}w_{j'}$$

By grouping all terms that give $w_1w_2w_3w_4$ we find

$$E[X_1X_2X_3X_4] = \frac{1}{8}[8E[X_1X_2]E[X_3X_4] + 8E[X_1X_3]E[X_2X_4] + 8E[X_1X_4]E[X_2X_3]]$$
$$= E[X_1X_2]E[X_3X_4] + E[X_1X_3]E[X_2X_4] + E[X_1X_4]E[X_2X_3] .$$

4.9 Mean Square Estimation

4.91 From Eqns. 4.100AB, letting D correspond to X_1 and B to X_2, the best coefficients are:

a)
$$a = \frac{\sigma_B^2 COV(D,E) - COV(B,D)COV(B,E)}{\sigma_D^2\sigma_B^2 - COV(B,D)^2}$$
$$= \frac{\sigma^2(\sigma^2\rho) - \sigma^2\rho^2\sigma^2\rho}{\sigma^4 - \sigma^4\rho^4} = \frac{\rho - \rho^3}{1-\rho^4} = \frac{\rho}{1+\rho^2}$$

and

$$b = \frac{\sigma_D^2 COV(B,E) - COV(B,D)COV(D,E)}{\sigma_D^2\sigma_B^2 - COV(B,D)^2}$$
$$= \frac{\sigma^2(\sigma^2\rho) - \sigma^2\rho^2(\sigma^2\rho)}{\sigma^4 - \sigma^4\rho^4} = \frac{\rho}{1+\rho^2}$$

b) The minimum mean square error is:

$$\mathcal{E}[(E - \underbrace{(aD+bB)}_{\hat{E}})^2] = \mathcal{E}[(E-\hat{E})(E-aD-bB)]$$
$$= \mathcal{E}[(E-\hat{E})E] - a\underbrace{\mathcal{E}[(E-\hat{E})D]}_{0} - b\underbrace{\mathcal{E}[(E-\hat{E})B]}_{0}$$

since error and observations are orthogonal
$$\begin{aligned}
&= \mathcal{E}[(E - aD - bB)E] \\
&= \mathcal{E}[E^2] - a\mathcal{E}[DE] - b\mathcal{E}[BE] \\
&= \sigma^2 - \frac{\rho}{1+\rho^2}\rho\sigma^2 - \frac{\rho}{1+\rho^2}\rho\sigma^2 \\
&= \sigma^2 - \frac{2\rho^2}{1+\rho^2}\sigma^2 = \sigma^2\left\{1 - \frac{2\rho^2}{1+\rho^2}\right\}
\end{aligned}$$

4.10 Generating Correlated Vector Random Variables

4.95 a) Let X_1 and X_2 be the zero-mean, unit-variance Gaussian RV's.
1. Generate a Gaussian RV with mean m_1 and variance σ_1^2, that is, let $Z_1 = \sigma_1 X_1 + m_1$.
2. Generate a Gaussian RV with conditional mean and variance given in (4.89)

$$\begin{aligned}
Z_2 &= \sqrt{1-\rho^2}\sigma_2 X_2 + \rho\frac{\sigma_2}{\sigma_1}(Z_1 - m_1) + m_2 \\
&= \sqrt{1-\rho^2}\sigma_2 X_2 + \sigma_2\rho X_1 + m_2
\end{aligned}$$

$$\therefore \begin{bmatrix} Z_1 \\ Z_2 \end{bmatrix} = \begin{bmatrix} \sigma_1 & 0 \\ \sigma_2\rho & \sigma_2\sqrt{1-\rho^2} \end{bmatrix} \begin{bmatrix} X_1 \\ X_2 \end{bmatrix} + \begin{bmatrix} m_1 \\ m_2 \end{bmatrix}$$

4.11 Problems Requiring Cumulative Knowledge

4.100 a) Let N be the number of items arrivals between inspections. N is a geometric random variable with pmf

$$P[X = k] = p(1-p)^{k-1} \quad k = 1, 2, \ldots$$

b) Let T_i be the time between arrivals of items $i-1$ and i. The time between inspected items is

$$X = \sum_{i=1}^{N} T_i$$

4.11. Problems Requiring Cumulative Knowledge

where N is the above geometric random variable. Using conditional expectation, we have

$$f_X(t) = \sum_{k=1}^{\infty} f_X(t|N=k) P[N=k]$$

When $N = k$, X is an Erlang random variable with pdf

$$f_X(t|N=k) = \frac{\lambda(\lambda t)^{k-1} e^{-\lambda t}}{(k-1)!}$$

Therefore

$$\begin{aligned}
f_X(t) &= \sum_{k=1}^{\infty} \frac{\lambda(\lambda t)^{k-1}}{(k-1)!} e^{-\lambda t} p(1-p)^{k-1} \\
&= \lambda p e^{-\lambda t} \sum_{k=1}^{\infty} \frac{(\lambda t(1-p))^{k-1}}{(k-1)!} \\
&= \lambda p e^{-\lambda t} e^{\lambda t(1-p)} \\
&= \lambda p e^{-\lambda p t}
\end{aligned}$$

Thus X is an exponential random variable with parameter λp. We can explain this result by recalling that the exponential random variable is a limiting form of the geometric random variable (page 112). In the original exponential random, the probability of success in a Bernoulli trial was proportional to λ. In our modified scenario, we flip an additional coin to determine whether a given success (arrival) is to be inspected. Thus we have a new sequence of Bernoulli trials in which the probability of success is proportional to λp. In the limit, this leads to the exponential random variable with parameter λp.

$$P[X < t] = 1 - e^{-\lambda p t} = 0.9$$

Thus

$$e^{-\lambda p t} = \frac{1}{10} \Rightarrow p = \frac{1}{\lambda t} \ln 10$$

4.105 a) Since $Y = X + N$ then $\mathcal{E}[Y] = \mathcal{E}[X] + \mathcal{E}[N] = 0$.
To determine ρ we need the following:

$$\begin{aligned}
\mathcal{E}[XY] &= \mathcal{E}[X(X+N)] = \mathcal{E}[X^2] + \mathcal{E}[X]\mathcal{E}[N] = \mathcal{E}[X^2] = \sigma_X^2 \\
VAR[Y] &= VAR[X+N] = VAR[X] + VAR[N] \\
&= \sigma_X^2 + \sigma_N^2 \quad \text{since } X \text{ and } N \text{ are uncorrelated} \\
&\quad \text{(see Prob. 4.50b)}
\end{aligned}$$

Thus

$$\rho = \frac{COV(X,Y)}{\sigma_X \sigma_Y} = \frac{\mathcal{E}[XY]}{\sigma_X \sigma_Y} = \frac{\sigma_X^2}{\sigma_X \sqrt{\sigma_X^2 + \sigma_N^2}} = \frac{1}{\sqrt{1 + \frac{\sigma_N^2}{\sigma_X^2}}}$$

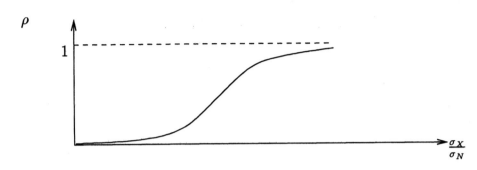

b) The best linear estimator for estimating the "signal" X from the "noisy signal" Y is:

$$\hat{X} = \rho_{XY} \frac{\sigma_X}{\sigma_Y} Y = \frac{\sigma_X^2}{\sigma_X^2 + \sigma_N^2} Y$$

The minimum mean square error is:

$$\begin{aligned}\mathcal{E}[(X-\hat{X})^2] &= \sigma_X^2(1-\rho^2) \\ &= \frac{\sigma_N^2}{\sigma_X^2 + \sigma_N^2}\sigma_X^2\end{aligned}$$

Chapter 5

Sums of Random Variables and Long-Term Averages

5.1 Sums of Random Variables

5.1 $\mathcal{E}[X+Y+Z] = \mathcal{E}[X] + \mathcal{E}[Y] + \mathcal{E}[Z] = 0$

a) From Eqn. 5.3 we have

$$\begin{aligned} VAR(X+Y+Z) &= VAR(X) + VAR(Y) + VAR(Z) \\ &\quad + 2COV(X,Y) + 2COV(X,Z) + 2COV(Y,Z) \\ &= 1+1+1+2\left(\frac{1}{4}\right) + 2(0) + 2\left(-\frac{1}{4}\right) = 3 \end{aligned}$$

b) From Eqn. 5.3 we have

$$VAR(X+Y+Z) = VAR(X) + VAR(Y) + VAR(Z) = 3$$

5.4 a) By Eqn. 5.7 we have

$$\Phi_Z(\omega) = \Phi_X(\omega)\Phi_Y(\omega) = e^{-\alpha|\omega|}e^{-\beta|\omega|} = e^{-(\alpha+\beta)|\omega|}$$

b) Taking the inverse transform:

$$f_Z(z) = \Phi_Z^{-1}(\omega) = \frac{1}{\lambda}\frac{\alpha+\beta}{(\alpha+\beta)^2 + z^2} \Rightarrow Z \quad \text{is also Cauchy}$$

5.10
$$G_{S_k}(z) = \mathcal{E}[z^{X_1+\ldots+X_k}] = \mathcal{E}[z^{X_1}]\ldots\mathcal{E}[z^{X_k}] = G_{X_1}(z)\ldots G_{X_k}(z)$$
$$= [pz+q]^{n_1}[pz+q]^{n_2}\ldots[pz+q]^{n_k}$$
$$= [pz+q]^{n_1+\ldots+n_k}$$

where the second equality follows from the independence of the X_i's. The result states that X_k is Binomial with parameters $n_1 + \ldots + n_k$ and p. This is obvious since S_k is the number of heads in $n_1 + \ldots + n_k$ tosses.

5.12 a) Note first that

$$\mathcal{E}[S/N=n] = \mathcal{E}\left[\sum_{k=1}^n X_k\right] = nE[X],$$

thus
$$\mathcal{E}[S] = \mathcal{E}[\mathcal{E}[S/N]] = \mathcal{E}[N\mathcal{E}[X]] = \mathcal{E}[N]\mathcal{E}[X], \text{ since } E[X] \text{ is a constant.}$$

Next consider
$$\mathcal{E}[S^2] = \mathcal{E}[\mathcal{E}[S^2/N]]$$

which requires that we find

$$\mathcal{E}[S^2|N=n] = \mathcal{E}\left[\sum_{i=1}^n X_i \sum_{j=1}^n X_j\right] = \sum_{i=1}^n \sum_{j=1}^n \mathcal{E}[X_i X_j]$$
$$= n\mathcal{E}[X^2] + n(n-1)\mathcal{E}[X]^2$$

since $E[X_i X_j] = \mathcal{E}[X^2]$ if $i=j$ and $E[X_i X_j] = \mathcal{E}[X]^2$ if $i \neq j$. Thus

$$\mathcal{E}[S^2] = \mathcal{E}[N\mathcal{E}[X^2] + N(N-1)\mathcal{E}[X]^2]$$
$$= \mathcal{E}[N]\mathcal{E}[X^2] + \mathcal{E}[N^2]\mathcal{E}[X^2] - \mathcal{E}[N]\mathcal{E}[X]^2$$

Then
$$VAR(S) = \mathcal{E}[S^2] - \mathcal{E}[S]^2$$
$$= \mathcal{E}[N]\mathcal{E}[X^2] + \mathcal{E}[N^2]\mathcal{E}[X^2] - \mathcal{E}[N]\mathcal{E}[X]^2 - \mathcal{E}[N]^2\mathcal{E}[X]^2$$
$$= \mathcal{E}[N]VAR[X] + VAR[N]\mathcal{E}[X]^2$$

b) First note that

$$\mathcal{E}[z^S/N=n] = \mathcal{E}[z^{\sum_{i=1}^n X_i}] = \mathcal{E}[z^{X_1}]...\mathcal{E}[z^{X_n}] = G_X(z)^n$$

Then

$$\begin{aligned}\mathcal{E}[z^S] &= \mathcal{E}[\mathcal{E}[z^S|N] \\ &= \mathcal{E}[G_X^N(z)] \\ &= \mathcal{E}[\omega^N]_{\omega=G_X(z)} \\ &= G_N(G_X(z))\end{aligned}$$

5.2 The Sample Mean and the Laws of Large Numbers

5.15
$$\begin{aligned}P\left[\left|\frac{N(t)}{t}-\lambda\right|\geq\varepsilon\right] &= P[|N(t)-\lambda t|\geq\varepsilon t] \\ &\leq \frac{VAR[N(t)]}{(\varepsilon t)^2} \quad \text{by Chebyshev Inq.} \\ &= \frac{\lambda t}{\varepsilon^2 t^2} = \frac{\lambda}{\varepsilon^2 t}\end{aligned}$$

5.18 For $n=10$, Eqn. 5.20 gives

$$P[|M_{10}-0|<\varepsilon] \geq 1-\frac{1^2}{10\varepsilon^2} = 1-\frac{1}{10}\frac{1}{\varepsilon^2}$$

Since M_{10} is Gaussian with mean 0 and variance $\frac{1}{10}$

$$\begin{aligned}P[|M_{10}-0|<\varepsilon] &= P[-\varepsilon<M_{10}<\varepsilon] = 1-2Q(\sqrt{10}\varepsilon) \\ &= 1-2Q(3.16\varepsilon)\end{aligned}$$

Similarly for $n=100$ we obtain

$$\begin{aligned}P[|M_{100}-0|<\varepsilon] &= 1-\frac{1}{100}\frac{1}{\varepsilon^2} \\ P[|M_{100}-0|<\varepsilon] &= 1-2Q(10\varepsilon)\end{aligned}$$

For example if $\varepsilon = \frac{1}{2}$

$$P[|M_{10}| < \frac{1}{2}] \geq 1 - \frac{1}{10/4} = .6$$

$$P[|M_{10}| < \frac{1}{2}] = 1 - 2Q(1.58) = 1 - 2(5.44(10^{-2})) = .89$$

$$P[|M_{100}| < \frac{1}{2}] \geq 1 - \frac{1}{100/4} = .96$$

$$P[|M_{100}| < \frac{1}{2}] = 1 - 2Q(5) = 1 - 2(2.87)10^{-7}$$

Note the significant discrepancies between the bounds and the exact values.

5.3 The Central Limit Theorem

5.22 The relevant parameters are $n = 1000$, $m = np = 500$, $\sigma^2 = npq = 250$. The Central Limit Theorem then gives:

$$P[400 \leq N \leq 600] = P\left[\frac{400 - 500}{\sqrt{250}} \leq \frac{N - m}{\sigma} \leq \frac{600 - 500}{\sqrt{250}}\right]$$
$$\approx Q(-6.324) - Q(6.324) = 1 - 2Q(6.324)$$
$$= 1 - 2.54(10^{-10})$$
$$P[500 \leq N \leq 550] \approx Q(0) - Q(3.162) = \frac{1}{2} - 7.3(10^{-4})$$

5.29 The total number of errors S_{100} is the sum of iid Bernoulli random variables

$$S_{100} = X_1 + ... + X_{100}$$
$$\mathcal{E}[S_{100}] = 100p = 15$$
$$VAR[S_{100}] = 100pq = 12.75$$

The Central Limit Theorem gives:

$$P[S_{100} \leq 20] = 1 - P[S_{100} > 20]$$
$$= 1 - P\left[\frac{S_{100} - 15}{\sqrt{12.75}} > \frac{20 - 15}{\sqrt{12.75}}\right]$$
$$\approx 1 - Q(1.4) = 0.92$$

5.4 Confidence Intervals

5.31 The ith measurement is $X_i = m + N_i$ where $\mathcal{E}[N_i] = 0$ and $VAR[N_i] = 10$. The sample mean is $M_{100} = 100$ and the variance is $\sigma = \sqrt{10}$.

Eqn. 5.37 with $z_{\alpha/2} = 1.96$ gives

$$\left(100 - \frac{1.96\sqrt{10}}{\sqrt{30}}, 100 + \frac{1.96\sqrt{10}}{\sqrt{30}}\right) = (98.9, 101.1)$$

5.37 The sample mean and variance of the batch sample means are $M_{10} = 24.9$ and $V_{10}^2 = 3.42$. The mean number of heads in a batch is $\mu = \mathcal{E}[M_{10}] = \mathcal{E}[X] = 50p$. From Table 5.2, with $1 - \alpha = 95\%$ and $n - 1 = 9$ we have

$$z_{\alpha/2,9} = 2.262$$

The confidence interval for μ is

$$\left(M_{10} - \frac{z_{\alpha/2,9} V_{10}}{\sqrt{10}}, M_{10} + \frac{z_{\alpha/2,9} V_{10}}{\sqrt{10}}\right) = (23.58, 26.22)$$

The confidence interval for $p = M_{10}/50$ is then

$$\left(\frac{23.58}{50}, \frac{26.22}{50}\right) = (0.4716, 0.5244)$$

5.5 Convergence of Sequences of Random Variables

5.40 We'll consider $U_n(\xi)$:

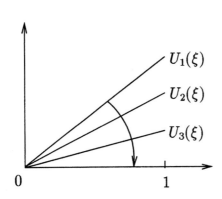
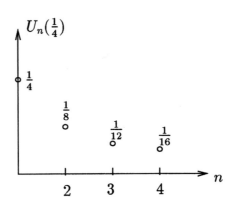

5.45 We are given that $X_n \to X$ ms and $Y_n \to Y$ ms. Consider

$$\mathcal{E}[((X_n + Y_n) - (X + Y))^2] = E[((X_n - X) + (Y_n - Y))^2]$$
$$= E[(X_n - X)^2] + E[(Y_n - Y)^2]$$
$$+ 2E[(X_n - X)(Y_n - Y)]$$

The first two terms approach zero since $X_n \to X$ and $Y_n \to Y$ in mean square sense. We need to show that the last term also goes to zero. This requires the Schwarz Inequality:

$$E[ZW] \leq \sqrt{E[Z^2]}\sqrt{E[W^2]}\ .$$

When the inequality is applied to the third term we have:

$$E[((X_n + Y_n) - (X + Y))^2] \leq E[(X_n - X)^2] + E[(Y_n - Y)^2]$$
$$+ 2\sqrt{E[(X_n - X)^2]}\sqrt{E[(Y_n - Y)^2]}$$
$$= (\sqrt{E[(X_n - X)^2]} + \sqrt{E[(Y_n - Y)^2]})^2$$
$$\to 0 \quad \text{as } n \to \infty\ .$$

To prove the Schwarz Inequality we take

$$0 \leq E[(Z + aW)^2]$$

and minimize with respect to a:

$$\frac{d}{da}(E[Z^2] + 2aE[ZW] + a^2 E[W^2]) = 0$$
$$2E[ZW] + 2aE[W^2] = 0$$

5.6. Long-Term Arrival Rates and Associated Averages

$\Rightarrow$ minimum attained by $a^* = -\dfrac{E[ZW]}{E[W^2]}$. Thus

$$0 \leq E[(Z + a^*W)^2] = E[Z^2] = 2\dfrac{E[ZW]^2}{E[W^2]} + \dfrac{E[ZW]^2}{E[W^2]}$$

$$\Rightarrow \dfrac{E[ZW]^2}{E[W^2]} \leq E[Z^2]$$

$$\Rightarrow E[ZW] \leq \sqrt{E[Z^2]}\sqrt{E[W^2]} \quad \text{as required}$$

5.6 Long-Term Arrival Rates and Associated Averages

5.51 Let Y be the bus interdeparture time, then

$$Y = X_1 + X_2 + \ldots + X_m \quad \text{and} \quad \mathcal{E}[Y] = m\mathcal{E}[X_i] = mT$$

$\therefore$ long-term bus departure rate $= \dfrac{1}{\mathcal{E}[Y]} = \dfrac{1}{mT}$.

5.53 Show $\{N(t) \geq n\} \Leftrightarrow \{S_n \leq t\}$.

a) We first show that $\{N(t) \geq n\} \Rightarrow \{S_n \leq t\}$.
If $\{N(t) \geq n\} \Rightarrow t \geq S_{N(t)} = X_1 + X_2 + \ldots + X_{N(t)}$
$\geq X_1 + \ldots + X_n = S_n$
$\Rightarrow \{S_n \leq t\}$

Next we show that $\{S_n \leq t\} \Rightarrow \{N(t) \geq n\}$. If $\{S_n \leq t\}$ then the nth event occurs before time t

$\Rightarrow N(t)$ is at least n
$\Rightarrow \{N(t) \geq n\} \quad \checkmark$

b)
$$\begin{aligned}P[N(t) \le n] &= 1 - P[N(t) \ge n+1] \\ &= 1 - P[S_{n+1} \le t] \\ &= 1 - \left(1 - \sum_{k=0}^{n} \frac{(\alpha t)^k}{k!} e^{-\alpha t}\right) \\ &= \sum_{k=0}^{n} \frac{(\alpha t)^k}{k!} e^{-\alpha t}\end{aligned}$$

since S_{n+1} is an Erlang RV. Thus $N(t)$ is a Poisson random variable.

5.59 The interreplacement time is

$$\tilde{X}_i = \begin{cases} X_i & \text{if } X_i < 3T \quad \text{that is, item breaks down before } 3T \\ 3T & \text{if } X_i \ge 3T \quad \text{that is, item is replaced at time } 3T \end{cases}$$

where the X_i are iid exponential random variables with mean $\mathcal{E}[X_i] = T$.

The mean of $\tilde{X}_i$ is:

$$\mathcal{E}[\tilde{X}_i] = \int_0^{3T} x \frac{1}{T} e^{-x/T} dx + 3T P[X > 3T] = T(1 - e^{-3})$$

a) Therefore the

$$\begin{array}{c}\text{long-term} \\ \text{replacement} \\ \text{rate}\end{array} = \frac{1}{\mathcal{E}[\tilde{X}]} = \frac{1}{T(1-e^{-3})}$$

b) Let

$$c_i = \begin{cases} 1 & X_i \ge 3T \\ 0 & X_i < 3T \end{cases} \quad \text{i.e., a good item is replaced}$$

Then

$$\mathcal{E}[C] = P[X_i \ge 3T] = e^{-3}$$

∴ long term rate at which working components are replaced is

$$\lim_{t \to \infty} \frac{\sum_{i=0}^{N(t)} C_i}{t} = \frac{\mathcal{E}[C]}{\mathcal{E}[\tilde{X}]} = \frac{e^{-3}}{T(1-e^{-3})}$$

5.6. Long-Term Arrival Rates and Associated Averages

5.63 a) Since the age $a(t)$ is the time that has elapsed from the last arrival up to time t, then

$$C_j = \int_0^{X_j} a(t')dt' = \int_0^{X_j} t'dt' = \frac{X_j^2}{2}$$

The figure below shows the relation between $a(t)$ and the C_j's.

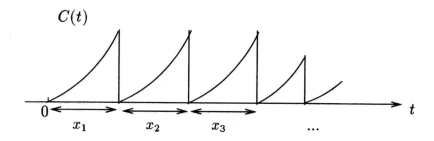

b) $\lim\limits_{t\to\infty} \dfrac{C(t)}{t} = \dfrac{\mathcal{E}[C]}{\mathcal{E}[X]} = \dfrac{\mathcal{E}[X^2]}{2\mathcal{E}[X]}$

c) From the above figure:

$$\lim_{t\to\infty} \frac{1}{t}\int_0^t a(t')dt' = \lim_{t\to\infty} \frac{1}{t}\sum_{j=1}^{N(t)} \int_0^{X_j} a(t')dt'$$

$$= \lim_{t\to\infty} \frac{1}{t}\sum_{j=1}^{N(t)} C_j$$

$$= \frac{\mathcal{E}[X^2]}{2\mathcal{E}[X]} \quad \text{from part b)}$$

d) For the residual life in a cycle

$$C'_j = \int_0^{X_j} r(t')dt' = \int_0^{X_j} (X_j - t')dt' = \frac{X_j^2}{2} = C_j$$

$\Rightarrow$ same cost as for age of a cycle

5.7 A Computer Method for Evaluating the Distribution of a Random Variable Using the Discrete Fourier Transform

5.66 a)
$$c_0 = \frac{1}{3} + \frac{1}{3} + \frac{1}{3} = 1$$
$$c_1 = \frac{1}{3} + \frac{1}{3}e^{j\frac{2\pi}{3}} + \frac{1}{3}e^{j\frac{2\pi}{3}} = 0$$
$$c_2 = \frac{1}{3} + \frac{1}{3}e^{j\frac{4\pi}{3}} + \frac{1}{3}e^{j\frac{8\pi}{3}} = 0$$

b) $P[X = 1] = \frac{1}{3}\left[1 + 0 \cdot e^{-j\frac{2}{3}\pi} + 0 \cdot e^{-j\frac{4}{3}\pi}\right] = \frac{1}{3}$

5.8 Problems Requiring Cumulative Knowledge

5.78 $X_n = \frac{1}{2}U_n + \left(\frac{1}{2}\right)^2 U_{n-1} + \ldots + \left(\frac{1}{2}\right)^n U_1 \quad n \geq 1$

This "low-pass filter" weighs recent samples more heavily than older samples. Note that we can also write X_n as follows:

$$X_n = \frac{1}{2}X_{n-1} + \frac{1}{2}U_n \quad X_0 = 0, \quad n \geq 1$$

We will see in Chapter 6 that X_n is an autoregressive random process.

a)
$$E[X_n] = E\left[\frac{1}{2}\sum_{j=0}^{n-1}\left(\frac{1}{2}\right)^j U_{n-j}\right] = \frac{1}{2}\sum_{j=0}^{n-1}\left(\frac{1}{2}\right)^j E[U]$$
$$= \frac{1}{2}E[U]\frac{1-(\frac{1}{2})^n}{1-\frac{1}{2}} = E[U]\left(1 - \left(\frac{1}{2}\right)^n\right)$$
$$= 0 \quad \text{since } E[U] = 0.$$

$$E[X_n^2] = E\left[\frac{1}{2}\sum_{j=0}^{n-1}\left(\frac{1}{2}\right)^n U_{n-j} \frac{1}{2}\sum_{j'=0}^{n-1}\left(\frac{1}{2}\right)^{j'} U_{n-j'}\right]$$
$$= \frac{1}{4}\sum_{j=0}^{n-1}\sum_{j'=0}^{n-1}\left(\frac{1}{2}\right)^{j+j'} E[U_{n-j}U_{n-j'}]$$

5.8. Problems Requiring Cumulative Knowledge

$$= \frac{1}{4}\sum_{j=0}^{n-1}\left(\frac{1}{2}\right)^{2j}E[U^2] \quad \text{since the } U_j \text{ are iid}$$

$$= \frac{\sigma^2}{3}\left(1-\left(\frac{1}{4}\right)^n\right) \quad \text{where } E[U^2]=\sigma^2$$

$$VAR(X_n) = E[X_n^2] - E[X_n]^2 = \frac{\sigma^2}{3}\left(1-\left(\frac{1}{4}\right)^n\right).$$

We could also obtain these results as follows:

$$E[X_n] = E\left[\frac{1}{2}X_{n-1}\right] + \frac{1}{2}E[U_n] \quad E[X_0] = 0$$

$$= \frac{1}{2}E[X_{n-1}] + \frac{1}{2}E[U] \quad n \geq 1$$

This first-order difference equation has solution

$$E[X_n] = E[U]\left(1-\left(\frac{1}{2}\right)^n\right).$$

For the second moment, we have:

$$E[X_n^2] = E\left[\left(\frac{1}{2}X_{n-1}+\frac{1}{2}U_n\right)^2\right]$$

$$= \frac{1}{4}E[X_{n-1}^2] + \underbrace{\frac{1}{2}e[X_{n-1}U_n]}_{0} + \frac{1}{4}E[U_n^2]$$

$$= \frac{1}{4}E[X_{n-1}^2] + \frac{1}{4}E[U^2] \quad \begin{array}{l} E[X_0^2]=0 \\ n \geq 1 \end{array}$$

This first-order difference equation has solution:

$$E[X_n^2] = \frac{\sigma^2}{3}\left(1-\left(\frac{1}{4}\right)^n\right).$$

b) The characteristic function of X_n is:

$$\Phi_{X_n}(\omega) = E[e^{j\omega X_n}] = E\left[e^{j\omega\frac{1}{2}\sum_{j=0}^{n-1}(\frac{1}{2})^n U_{n-j}}\right]$$

$$= E\left[e^{j\frac{\omega}{2}U_n}\right]E\left[e^{j\frac{\omega}{4}U_{n-1}}\right]\ldots E\left[e^{j\frac{\omega}{2^n}U_1}\right]$$

U_n are iid Gaussian KV's with characteristic function

$$\Phi_{U_n}(\omega) = E[e^{j\omega U_n}] = e^{-\frac{1}{2}\sigma^2\omega^2}$$

Therefore

$$\begin{aligned}
\Phi_{X_n}(\omega) &= e^{-\frac{1}{2}\sigma^2(\frac{\omega}{2})^2} e^{-\frac{1}{2}\sigma^2(\frac{\omega}{4})^2} \cdots e^{-\frac{1}{2}\sigma^2(\frac{\omega}{2^n})^2} \\
&= e^{-\frac{1}{2}\sigma^2\omega^2(\frac{1}{4}+(\frac{1}{4})^2+\cdots(\frac{1}{4})^n)} \\
&= e^{-\frac{1}{2}\frac{\sigma^2}{3}(1-(\frac{1}{4})^n)\omega^2}
\end{aligned}$$

Thus X_n is a zero-mean Gaussian random variable with the variance found in part a).

As $n \to \infty$

$$\Phi_{X_n}(\omega) \to e^{-\frac{1}{2}\frac{\sigma^2}{3}\omega^2}$$

so X_n approaches a zero-mean Gaussian random variable with variance $\sigma^2/3$.

c) The result in Part b) shows that X_n converges in distribution to a Gaussian random variable X with zero mean and variance $\sigma^2/3$.

To determine whether X_n converges in mean-square sense consider the Cauchy Criterion in Eq. 5.50. Consider X_n and X_{n+m}:

$$\begin{aligned}
\mathcal{E}[(X_{n+m} - X_n)^2] &= \mathcal{E}\left[\left(\frac{1}{2}\sum_{j=0}^{n+m-1}\left(\frac{1}{2}\right)^n U_{n-j} - \frac{1}{2}\sum_{j'=0}^{n+m-1}\left(\frac{1}{2}\right)^{j'} U_{n-j'}\right)^2\right] \\
&= \frac{1}{4}\mathcal{E}\left[\left(\sum_{j=n}^{n+m-1}\left(\frac{1}{2}\right)^j U_{n-j}\right)^2\right] \\
&= \frac{1}{4}\sum_{j=n}^{n+m-1}\sum_{j'=n}^{n+m-1}\left(\frac{1}{2}\right)^{j+j'} E[U_{n-j}U_{n-j'}] \\
&= \frac{\sigma^2}{4}\sum_{j=n}^{n+m-1}\left(\frac{1}{2}\right)^{2j} \\
&= \frac{\sigma^2}{4}\left(\frac{1}{4}\right)^n \left(\frac{1-(\frac{1}{4})^m}{1-\frac{1}{4}}\right) \\
&\to 0 \quad \text{as } n,m \to \infty
\end{aligned}$$

Therefore X_n converges in mean-square sense.

To determine almost-sure convergence of X_n would take us beyond the scope of the text. See Gray and Davisson, page 183 for a discussion on how this is done.

Chapter 6

Random Processes

6.1 Definition and Specification of a Stochastic Process

6.1 We find the probabilities of the events $\{X_1 = i, X_2 = j\}$ in terms of the probabilities of the equivalent events of ξ:

$$P[X_1 = 1, X_2 = 1] = P[\frac{3}{4} < \xi < 1] = \frac{1}{4}$$
$$P[X_1 = 0, X_2 = 1] = P[\frac{1}{4} < \xi < \frac{1}{2}] = \frac{1}{4}$$
$$P[X_1 = 1, X_2 = 0] = P[\frac{1}{2} < \xi < \frac{3}{4}] = \frac{1}{4}$$
$$P[X_1 = 0, X_2 = 0] = P[0 < \xi < \frac{1}{4}] = \frac{1}{4}$$

$\Rightarrow P[X_1 = i, X_2 = j] = P[X_1 = i]P[X_2 = j]$ all $i, j \in \{0, 1\}$
$\Rightarrow X_1, X_2$ independent RV's

6.5 a) Since $g(t)$ is zero outside the interval [0,1]:

$$P[X(t) = 0] = 1 \text{ for } t \notin [0, 1]$$

For $t \in [0, 1]$, we have

$$P[X(t) = 1] = P[X(t) = -1] = \frac{1}{2}$$

73

b) $m_X(t) = \begin{cases} 1 \cdot P[X(t) = 1] + (-1)P[X(t) = -1] = 0 & 0 \leq t \leq 1 \\ 0 & \text{otherwise} \end{cases}$

c) For $t \in [0,1]$, $t + d \in [0,1]$, $X(t)$ must be the same value, thus:

$$P[X(t) = \pm 1, X(t+d) = \pm 1] = \frac{1}{2}$$
$$P[X(t) = \pm 1, X(t+d) = \mp 1] = 0$$

For $t \in [0,1]$, $t + d \notin [0,1]$:

$$P[X(t) = \pm 1, X(t+d) = 0] = \frac{1}{2}$$

For $t \notin [0,1]$, $t + d \notin [0,1]$:

$$P[X(t) = 0, X(t+d) = 0] = 1$$

d)
$$\begin{aligned} C_X(t, t+d) &= \mathcal{E}[X(t)X(t+d)] - m_X(t)m_X(t+d) \\ &= \mathcal{E}[X(t)X(t+d)] \\ &= \begin{cases} 1 & t \in [0,1] \text{ and } t+d \in [0,1] \\ 0 & \text{otherwise} \end{cases} \end{aligned}$$

6.7 a) We will use conditional probability:

$$\begin{aligned} P[X(t) \leq x] &= P[g(t-T) \leq x] \\ &= \int_0^1 P[g(t-T) \leq x | T = \lambda] f_T(\lambda) d\lambda \\ &= \int_0^1 P[g(t-\lambda) \leq x] d\lambda \quad \text{since } f_T(\lambda) = 1 \\ &= \int_{t-1}^t P[g(u) \leq x] du \quad \text{after letting } u = t - \lambda \end{aligned}$$

$g(u)$ (and hence $P[g(u) \leq x]$) is a periodic function of u with period 1, so we can change the limits of the above integral to any full period. Thus

$$P[X(t) \leq x] = \int_0^1 P[g(u) \leq x] du$$

Note that $g(u)$ is deterministic, so

$$P[g(u) \leq x] = \begin{cases} 1 & u : g(u) \leq x \\ 0 & u : g(u) > x \end{cases}$$

6.1. Definition and Specification of a Stochastic Process

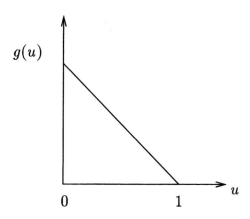

So finally
$$P[X(t) \leq x] = \int_{u:g(u)\leq x} 1 \, du = \int_{1-x}^{1} 1 \, du = x.$$

b) $m_X(t) = E[X(t)] = \int_0^1 x \, dx = \frac{1}{2}.$

The correlation is again found using conditioning on T:

$$\begin{aligned}
E[X(t)X(t+\tau)] &= \int_0^1 E[g(t-T)g(t+\tau-T)|T=\lambda] f_T(\lambda) d\lambda \\
&= \int_0^1 g(t-\lambda)g(t+\tau-\lambda) d\lambda \\
&= \int_{t-1}^{t} g(u)g(u+\tau) du
\end{aligned}$$

$g(u)g(u+\tau)$ is a periodic function in u so we can change the limits to $(0,1)$:

$$E[X(t)X(t+\tau)] = \int_0^1 g(u)g(u+\tau) du$$

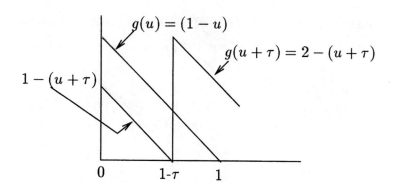

here we assume $0 < \tau < 1$ since $E[X(t)X(t+\tau)]$ is periodic in τ.

$$\begin{aligned} E[X(t)X(t+\tau)] &= \int_0^{1-\tau}(1-u)(1-u-\tau)du + \int_{1-\tau}^1(1-u)(2-u-\tau)du \\ &= \frac{1}{3} - \frac{\tau}{2} + \frac{\tau^3}{6} + \frac{\tau^2}{2} - \frac{\tau^3}{6} \\ &= \frac{1}{3} - \frac{\tau}{2} + \frac{\tau^2}{2} \: . \end{aligned}$$

Thus

$$\begin{aligned} C_X(t,t+\tau) &= \frac{1}{3} - \frac{\tau}{2} + \frac{\tau^2}{2} - \frac{1}{4} \\ &= \frac{1}{12} - \frac{\tau}{2} + \frac{\tau^2}{2} \end{aligned}$$

6.10 a) $P[H(t) = 1] = P[X(t) \geq 0] = P[\xi \cos 2\pi t \geq 0] = \frac{1}{2} = P[H(t) = -1]$

$$\begin{aligned} \mathcal{E}[H(t)] &= 1 \cdot P[H(t) = 1] + (-1)P[H(t) = -1] = 0 \\ C_H(t,t+\tau) &= \mathcal{E}[H(t)H(t+\tau)] \\ &= 1 \cdot P[\underbrace{H(t)H(t+\tau) = 1}_{\substack{H(t) \: \& \: H(t+\tau) \\ \text{same sign}}}] + (-1)P[\underbrace{H(t)H(t+\tau) = -1}_{\substack{H(t) \: \& \: H(t+\tau) \\ \text{opposite sign}}}] \end{aligned}$$

$$\begin{aligned} H(t)H(t+\tau) &= 1 \Leftrightarrow \cos 2\pi t \text{ and } \cos 2\pi(t+\tau) \text{ have same sign} \\ H(t)H(t+\tau) &= -1 \Leftrightarrow \cos 2\pi t \text{ and } \cos 2\pi(t+\tau) \text{ have different sign} \end{aligned}$$

$$\therefore C_H(t,t+\tau) = \begin{cases} 1 & \text{for } t,\tau \text{ such that } \cos 2\pi t \cos 2\pi(t+\tau) = 1 \\ -1 & \text{for } t,\tau \text{ such that } \cos 2\pi t \cos 2\pi(t+\tau) = -1 \end{cases}$$

b) $P[H(t) = 1] = P[X(t) \geq 0] = P[\cos(\omega t + \Theta) \geq 0] = \frac{1}{2} = P[H(t) = -1]$

6.1. Definition and Specification of a Stochastic Process

$$\mathcal{E}[H(t)] = 1\left(\frac{1}{2}\right) + (-1)\frac{1}{2} = 0$$

$$\mathcal{E}[H(t)H(t+\tau)] = 1 \cdot \underbrace{P[X(t)X(t+\tau) > 0]}_{1-P[X(t)X(t+\tau)<0]} + (-1)P[X(t)X(t+\tau) < 0]$$

$$= 1 - 2P[X(t)X(t+\tau) < 0]$$

$$\begin{aligned}
P[X(t)X(t+\tau) < 0] &= P[\cos(\omega t + \Theta)\cos(\omega(t+\tau) + \Theta) < 0] \\
&= \left[\frac{1}{2}\cos\omega\tau + \frac{1}{2}\cos(2\omega t + \omega\tau + 2\Theta) < 0\right] \\
&= P[\cos(2\omega t + \omega\tau + 2\Theta) < \cos\omega\tau] \\
&= 1 - \frac{\text{shaded region in figure}}{2\pi}
\end{aligned}$$

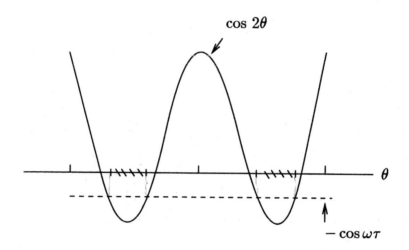

c) $P[H(t) = 1] = P[X(t) \geq 0] = 1 - F_{X(t)}(0^-) = 1 - P[H(t) = -1]$

$$\begin{aligned}
\mathcal{E}[H(t)] &= 1 \cdot P[H(t) = 1] + (-1)P[H(t) = -1] \\
&= 1 - F_{X(t)}(0^-) - F_{X(t)}(0^-) \\
&= 1 - 2F_{X(t)}(0^-)
\end{aligned}$$

d) $\mathcal{E}[H(t)X(t)] = \mathcal{E}[|X(t)|])$

since
$$H(t)X(t) = \begin{cases} +X(t) & X(t) \geq 0 \\ -X(t) & X(t) < 0. \end{cases}$$

6.14 The covariance matrix is given by:
$$K = \begin{bmatrix} C_X(t,t) & C_X(t,t+s) \\ C_X(t+s,t) & C_X(t+s,t+s) \end{bmatrix} = \begin{bmatrix} \sigma^2 & e^{-|s|} \\ e^{-|s|} & \sigma^2 \end{bmatrix}$$

Thus
$$|K|^{1/2} = [\sigma^2\sigma^2 - \sigma^2 e^{-|s|}\sigma^2 e^{-|s|}]^{1/2} = \sqrt{\sigma^4 - \sigma^4 e^{-2|s|}}$$
$$= \sigma^2\sqrt{1 - e^{-2|s|}}$$

and
$$K^{-1} = \frac{1}{\sigma^4(1 - e^{-2|s|})}\begin{bmatrix} \sigma^2 & -\sigma^2 e^{-|s|} \\ -\sigma^2 e^{-|s|} & \sigma^2 \end{bmatrix}$$
$$= \begin{bmatrix} \frac{1}{\sigma^2(1-e^{-2|s|})} & \frac{-e^{-|s|}}{\sigma^2(1-e^{-2|s|})} \\ \frac{-e^{-|s|}}{\sigma^2(1-e^{-2|s|})} & \frac{1}{\sigma^2(1-e^{-2|s|})} \end{bmatrix}$$

Thus the joint pdf is:
$$f_{X(t)X(t+s)}(x_1, x_2) = \frac{e^{-\frac{1}{2}\underline{x}'K^{-1}\underline{x}}}{2\pi\sigma^2\sqrt{1-e^{-2|s|}}} \quad \underline{x} = \begin{bmatrix} x_1 \\ x_2 \end{bmatrix}$$
$$= \frac{\exp\left\{-\frac{x_1^2 - 2e^{-|s|}x_1x_2 + x_2^2}{2(1-e^{-2|s|})\sigma^2}\right\}}{2\pi\sigma^2\sqrt{1-e^{-2|s|}}}$$

6.2 Examples of Discrete-Time Random Processes

6.21 a) Assume $n' > n$, $i \geq j$

$$P[S_n = j, S_{n'} = i] = P[S_n = j, \overbrace{S_{n'-n} = i - j}^{\text{increment}}]$$
$$= P[S_n = j]P[S_{n'-n} = i - j]$$
by indep. increment property

6.2. Examples of Discrete-Time Random Processes

In general

$$P[S_{n'} = i] \neq P[S_{n'-n} = i - j]$$
$$\therefore P[S_n = j, S_{n'} = i] \neq P[S_n = j]P[S_{n'} = i].$$

b) $P[S_{n_2} = j | S_{n_1} = i] = \dfrac{P[S_{n_2} = j, S_{n_1} = i]}{P[S_{n_1} = i]}$

$$= \dfrac{\overbrace{P[S_{n_2} - S_{n_1} = j - i]}^{\text{increment}} P[S_{n_1} = i]}{P[S_{n_1} = i]}$$

$$= P[S_{n_2} - S_{n_1} = j - i] = \binom{n_2 - n_1}{j - i} p^{j-i}(1-p)^{n_2 - n_1 - j + i}$$

c.)

$$\begin{aligned}
P[S_{n_2} | S_{n_1} = i, S_{n_0} = k] &= \dfrac{P[S_{n_2} = j, S_{n_1} = i, S_{n_0} = k]}{P[S_{n_1} = i, S_{n_0} = k]} \\
&= \dfrac{P[S_{n_0} = k, S_{n_1} - S_{n_0} = i - k, S_{n_2} - S_{n_1} = j - i]}{P[S_{n_0} = k, S_{n_1} - S_{n_0} = i]} \\
&= \dfrac{P[S_{n_0} = k]P[S_{n_1} - S_{n_0} = i - k]P[S_{n_2} - S_{n_1} = j - i]}{P[S_{n_0} = k]P[S_{n_1} - S_{n_0} = i]} \\
&= P[S_{n_2} - S_{n_1} = j - i] \\
&= P[S_{n_2} = j | S_{n_1} = i].
\end{aligned}$$

6.26 Y_n and Z_n are Gaussian random processes with mean

$$\begin{aligned}
\mathcal{E}[Y_n] &= \tfrac{1}{2}\mathcal{E}[X_n] + \tfrac{1}{2}\mathcal{E}[X_{n-1}] = 0 \\
\mathcal{E}[Z_n] &= \tfrac{2}{3}\mathcal{E}[X_n] + \tfrac{1}{3}\mathcal{E}[X_{n-1}] = 0
\end{aligned}$$

and variance

$$\begin{aligned}
\mathcal{E}[Y_n^2] &= \mathcal{E}\left[\tfrac{1}{4}(X_n + X_{n-1}^2)\right] = \tfrac{1}{4}(1+1) = \tfrac{1}{2} \\
\mathcal{E}[Z_n^2] &= \mathcal{E}\left[\left(\tfrac{2}{3}X_n + \tfrac{1}{3}X_{n-1}\right)^2\right] = \tfrac{5}{9}
\end{aligned}$$

$$\therefore f_{Y_n}(y) = \dfrac{e^{-y^2}}{\sqrt{\pi}} \qquad f_{Z_n}(z) = \dfrac{3}{\sqrt{10\pi}}e^{-9z^2/10}$$

6.3 Examples of Continuous-Time Random Processes

6.34 Let $X_i =$ time till first arrival in line i

a) $$\begin{aligned} P[X_1 < X_2] &= \int_0^\infty P[X_2 > x | X_1 = x] f_{X_1}(x) dx \\ &= \int_0^\infty e^{-\lambda_2 x} \lambda_1 e^{-\lambda_1 x} dx = \lambda_1 \int_0^\infty e^{-(\lambda_1 + \lambda_2) x} dx \\ &= \frac{\lambda_1}{\lambda_1 + \lambda_2} \end{aligned}$$

b) Time till first arrival $\triangleq Z = \min(X_1, X_2)$

$$\begin{aligned} P[\min(X_1, X_2) > X] &= P[X_1 > x, X_2 > x] = P[X_1 > x] P[X_2 > x] \\ &= e^{-\lambda_1 x} e^{-\lambda_2 x} = e^{-(\lambda_1 + \lambda_2) x} \end{aligned}$$

$\Rightarrow Z$ exponential RV with mean $(\lambda_1 + \lambda_2)^{-1}$

c) $$\begin{aligned} G_{N(t)}(z) &= \mathcal{E}[z^{N_1(t) + N_2(t)}] = \mathcal{E}[z^{N_1(t)}] \mathcal{E}[z^{N_2(t)}] \\ &= e^{\lambda_1 (z-1)} e^{\lambda_2 (z-1)} = e^{(\lambda_1 + \lambda_2)(z-1)} \end{aligned}$$

$\Rightarrow N(t)$ Poisson with rate $\lambda_1 + \lambda_2$

d) $G_{N(t)} = e^{(\lambda_1 + \lambda_2 + \ldots + \lambda_n)(z-1)}$

6.39 Let $X_i = \#$ drinks dispensed to ith customer, then

$$X(t) = \sum_{i=1}^{N(t)} X_i$$

a) $\sum_{i=1}^n X_i$ is Poisson RV with mean n from Prob. 5.28.

$\Rightarrow P[X(t) = j | N(t) = n] = \frac{n^j}{j!} e^{-n}$

6.3. Examples of Continuous-Time Random Processes

b)
$$P[X(t) = j] = \sum_{n=0}^{\infty} \frac{n^j}{j!} e^{-n} \frac{(\lambda t)^n}{n!} e^{-\lambda t}$$

$$= \frac{e^{-\lambda t}}{j!} e^{\lambda t k} \sum_{n=0}^{\infty} n^j \frac{\left(\frac{\lambda t}{e}\right)^n}{n!} e^{-\lambda t/e}$$

$$= \frac{e^{-\lambda t(1-\frac{1}{e})}}{j!} \mathcal{E}[N^j]$$

where N Poisson RV with rate $\lambda t/e$

6.42 a) $P[Z(t) = 0 | Z(0) = 0] =$

$$P[\text{even \# transitions in } [0,t]] = \sum_{j=0}^{\infty} \frac{1}{1+\alpha t}\left(\frac{\alpha t}{1+\alpha t}\right)^{2j}$$

$$= \frac{1}{1+\alpha t} \frac{1}{1-\left(\frac{\alpha t}{1+\alpha t}\right)^2} = \frac{1+\alpha t}{1+2\alpha t}$$

$P[Z(t) = 0 | Z(0) = 1] =$

$$P[\text{odd \# transitions in } [0,t]] = \sum_{j=0}^{\infty} \frac{1}{1+\alpha t}\left(\frac{\alpha t}{1+\alpha t}\right)^{2j+1} = \frac{\alpha t}{1+2\alpha t}$$

$$P[Z(t) = 0] = P[Z(t) = 0 | Z(t) = 0]P[Z(0) = 0] + P[Z(t) = 0 | Z(0) = 1]P[Z(0) = 1]$$
$$= \frac{1+\alpha t}{1+2\alpha t}\frac{1}{2} + \frac{\alpha t}{1+2\alpha t}\frac{1}{2} = \frac{1}{2}$$

where we assume $P[Z(0) = 0] = \frac{1}{2}$

$$P[Z(t) = 1] = 1 - P[Z(t) = 0] = \frac{1}{2}$$

b) $m_Z(t) = 1 \cdot P[Z(t) = 1] = \frac{1}{2}$

6.46 a)
$$F_{Y(t)}(y) = P[Y(t) \leq y] = P[X(t) + \mu t \leq y] = P[X(t) \leq y - \mu t]$$
$$= F_{X(t)}(y - \mu t)$$

$$\Rightarrow f_{Y(t)}(y) = F_X(y - \mu t) = \frac{1}{\sqrt{2\pi \alpha t}} e^{-(y-\mu t)^2/2\alpha t}$$

b) $$\begin{aligned} F_{Y(t)Y(t+s)}(y_1,y_2) &= P[X(t)+\mu t \le y_1, X(t+s)+\mu(t+s) \le y_2] \\ &= F_{X(t),X(t+s)}(y_1-\mu t, y_2-\mu(t+s)) \\ \Rightarrow f_{Y(t)Y(t+s)}(y_1,y_2) &= f_{X(t),X(t+s)}(y_1-\mu t, y_2-\mu(t+s)) \\ &= f_{X(t)}(y_1-\mu t)f_{X(s)}(y_2-y_1-\mu s) \\ &= \frac{3^{-(y_1-\mu t)^2/2\alpha t}}{\sqrt{2\pi\alpha t}} \frac{e^{-(y_2-y_1-\mu s)^2/2\alpha s}}{\sqrt{2\pi\alpha s}} \end{aligned}$$

6.4 Stationary Random Process

6.51 $X(t) = \cos(\omega t + \Theta)$
From Example 6.7 $m_X(t) = 0$, $C_X(t_1,t_2) = \frac{1}{2}\cos\omega(t_1-t_2)$

$$\Rightarrow X(t) \quad \text{is wide sense stationary}$$

In order to determine whether $X(t)$ is stationary, consider the third-order joint pdf:

$$\begin{aligned} &f_{X(t_1)X(t_2)X(t_3)}(x_1,x_2,x_3)dx_1dx_2dx_3 \\ &= P[x_1 < \cos(\omega t_1+\Theta) \le x_1+dx_1, \; x_2 < \cos(\omega t_2+\Theta) \le x_2+dx_2, \\ &\qquad x_3 < \cos(\omega t_3+\Theta) \le x_3+dx_3] \\ &= P[A_1 \cap A_2 \cap A_3] \end{aligned}$$

where
$$A_i = \left\{ \cos^{-1}x_i - \omega t_i < \Theta \le \cos^{-1}x_i - \omega t_i + \frac{dx_i}{\sqrt{1-x_i^2}} \right\}$$

see Example 3.28 and Figure 3.19.

$$\begin{aligned} &f_{X(t_1+\tau)X(t_2+\tau)X(t_3+\tau)}(x_1,x_2,x_3) \\ &P[x_1 < \cos(\omega t_1+\omega\tau+\Theta) \le x_1+dx_1, \\ &\qquad x_2 < \cos(\omega t_2+\omega\tau+\Theta) \le x_2+dx_2, \\ &\qquad x_3 < \cos(\omega t_3+\omega\tau+\Theta) \le x_3+dx_3] \\ &= P[A_1' \cap A_2' \cap A_3'] \end{aligned}$$

where
$$A_i' = \left\{ \cos^{-1}x_i - \omega t_i - \omega\tau < \Theta \le \cos^{-1}x_i - \omega t_i - \omega\tau + \frac{dx_i}{\sqrt{1-x_i^2}} \right\}$$

6.4. Stationary Random Process

Since Θ is uniformly distributed, $P[A_i] = P[A_i']$.

In addition
$$P[A_1 \cap A_2 \cap A_3] = P[A_1' \cap A_2' \cap A_3']$$
since the intersection depends only on the relative values of t_1, t_2 and t_3. The same procedure can be used for nth order pdf's.

$\therefore X(t)$ is a <u>stationary</u> random process.

6.54 Assume X_n is discrete-valued, for simplicity, so that we can work with pmf's. Consider the third-order joint pmf of Y_n: for $n_1 < n_2 < n_3$ we need to show that for all $\tau > 0$

$$(\star) \quad P[Y_{n_1} = y_1, Y_{n_2} = y_2, Y_{n_3} = y_3] = P[Y_{n_1+\tau} = y_1, Y_{n_2+\tau} = y_2, Y_{n_3+\tau} = y_3]$$

Express the above probabilities in terms of the X_n's:

$$P[Y_{n_1} = y_1, Y_{n_2} = y_2, Y_{n_3} = y_3]$$
$$= P\left[\frac{1}{2}(X_{n_1} + X_{n_1-1}) = y_1, \frac{1}{2}(X_{n_2} + X_{n_2-1}) = y_2, \frac{1}{2}(X_{n_3} + X_{n_3-1}) = y_3\right]$$
$$= P\left[\frac{1}{2}(X_2 + X_1) = y_1, \frac{1}{2}(X_{n_2-n_1+2} + X_{n_2-n_1+1}) = y_2,\right.$$
$$\left.\frac{1}{2}(X_{n_3-n_1+2} + X_{n_3-n_1+1}) = y_3\right]$$

Since the joint pdf of $(X_{n_1-1}, X_{n_1}, X_{n_2-1}, X_{n_2}, X_{n_3-1}, X_{n_3})$ is identical to that of $(X_1, X_2, X_{n_2-n_1+1}, X_{n_2-n_1+2}, ..., X_{n_3-n_1+2})$ if X_n is a stationary process.

Similarly we have that

$$P[Y_{n_1+\tau} = y_1, Y_{n_2+\tau} = y_2, Y_{n_3+\tau} = y_3]$$
$$= P\left[\frac{1}{2}(X_{n_1+\tau} + X_{n_1+\tau-1}) = y_1, ... \frac{1}{2}(X_{n_3+\tau} + X_{n_3+\tau-1}) = y_3\right]$$
$$= P\left[\frac{1}{2}(X_2 + X_1) = y_1, \frac{1}{2}(X_{n_2-n_1+2} + X_{n_2-n_1+1}) = y_2, \frac{1}{2}(X_{n_3-n_1+2} + X_{n_3-n_1+1}) = y_3\right]$$

$\therefore (\star)$ holds if X_n is a stationary random process and in particular if X_n is an iid process.

6.57 a) $Z(t) = aX(t) + bY(t)$ so

$$\begin{aligned}
\mathcal{E}[Z(t)] &= a\mathcal{E}[X(t)] + b\mathcal{E}[Y(t)] = 0 \\
C_Z(t_1, t_2) &= \mathcal{E}[Z(t_1)Z(t_2)] = \mathcal{E}[(aX(t_1) + bY(t_1))(aX(t_2) + bY(t_2))] \\
&= a^2 \mathcal{E}[X(t_1)X(t_2)] + ab\mathcal{E}[X(t_1)Y(t_2)] \\
&\quad + ab\mathcal{E}[X(t_2)Y(t_1)] + b^2 \mathcal{E}[Y(t_1)Y(t_2)] \\
&= (a^2 + b^2)C_X(t_2 - t_1) \quad \text{since } \mathcal{E}[X(t_1)Y(t_2)] = 0
\end{aligned}$$

$\Rightarrow Z(t)$ is WSS.

b) Since $X(t)$ and $Y(t)$ are independent Gaussian RV's, $Z(t)$ is a Gaussian RV with mean 0 and variance $(a^2 + b^2)C_X(0)$.

$$f_{Z(t)}(x) = \frac{e^{-x^2/2(a^2+b^2)C_X(0)}}{\sqrt{2\pi(a^2+b^2)C_X(0)}}$$

6.5 Time Average of Random Processes and Ergodic Theorems

6.61 The sequence U_n and the sequences X_n, Y_n, and Z_n are related as shown below:

$$\begin{array}{cccccccccc}
\ldots & U_{-2} & U_{-1} & U_0 & U_1 & U_2 & U_3 & U_4 & U_5 & U_6 & \ldots \\
\ldots & Y_{-1} & Z_{-1} & X_0 & Y_0 & Z_0 & X_1 & Y_1 & Z_1 & X_2 & \ldots
\end{array}$$

If the sequence U_n is shifted by a multiple of 3, say be $3k$, then the subsequences X_n, Y_n and Z_n are shifted by k. If the subsequences are stationary, then the shifted U_n has the same joint distribution and thus U_n is <u>cyclostationary</u>.

On the other hand, if the shift of U_n is <u>not</u> a multiple of 3 then the joint distributions are not the same, and thus U_n is not stationary.

6.64 Recall the application of the Schwarz Inequality (Eqn. 6.60) during the discussion after the mean square periodic process was defined on page 360. We had:

$$E[(X(t+\tau+d) - X(t+\tau)X(t)]^2 \le E[(X(t+\tau+d) - X(t+\tau))^2]E[X^2(t)]$$

6.5. Time Average of Random Processes and Ergodic Theorems

If $X(t)$ is mean-square periodic, then

$$E[(X(t+\tau+d) - X(t+d))^2] = 0 \ .$$

Thus

$$E[(X(t+\tau+d) - X(t+\tau))X(t)]^2 = 0$$

$$\Rightarrow (E[X(t+\tau+d)X(t)] - E[X(t+\tau)X(t)])^2 = 0$$
$$\Rightarrow E[X(t+\tau+d)X(t)] = E[X(t+\tau)X(t)]$$
$$\Rightarrow R_X(t_1+d, t_2) = R_X(t_1, t_2)$$

Repeated applications of this argument to t_1 and t_2 imply

$$R_X(t_1 + md, t_2 + nd) = R_X(t_1, t_2) \quad \text{for every integer } m, n \ .$$

The special case $m = n$ implies (6.63b) and hence that $X(t)$ is wide-sense cyclostationary.

6.69

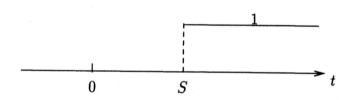

a) $P[X(t) \text{ discontinuous at } t_0] = P[s = t_0] = 0$

b)

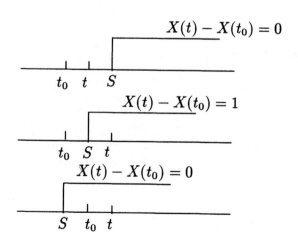

$$\lim_{t_0 \to 0} E[X(t) - X(t_0))^2] = 1 \cdot P[t_0 < S < t]$$
$$= e^{-\lambda t} - e^{-\lambda t_0}$$
$$\to 0 \quad \Rightarrow \quad X(t) \text{ is m.s. continuous}$$

We can also determine continuity from the autocorrelation function:

$$E[X(t_1)X(t_2)] = P[X > \max(t_1, t_2)]$$
$$= e^{-\lambda \max(t_1, t_2)}$$

Next we determine if $R_X(t_1, t_2)$ is continuous at (t_0, t_0):

$$R_X(t_0 + \varepsilon_1, t_0 + \varepsilon_2) - R_X(t_0, t_0) = e^{-\lambda \max(t_0 + \varepsilon_2, t_0 + \varepsilon_2)} - e^{-\lambda t_0}$$
$$= e^{-\lambda(t_0 + \max(\varepsilon_1, \varepsilon_2))} - e^{-\lambda t_0}$$
$$= e^{-\lambda t_0}[e^{-\lambda \max(\varepsilon_1, \varepsilon_2)} - 1]$$
$$\to 0 \quad \text{as } \varepsilon_1 \text{ and } \varepsilon_2 \to 0.$$
$$\Rightarrow \quad X(t) \text{ is m.s. continuous}$$

c) We expect that the mean square derivative is zero (if it exists). We thus consider the limit:

$$E\left[\left(\frac{X(t+\tau) - X(t)}{\varepsilon}\right)^2\right] = \frac{1}{\varepsilon^2}(e^{-\lambda(t+\varepsilon)} - e^{-\lambda t})$$
$$= e^{-\lambda t} \underbrace{\left(\frac{e^{-\lambda \varepsilon} - 1}{\varepsilon^2}\right)}_{\frac{\lambda \varepsilon - \frac{\lambda^2 \varepsilon^2}{2} + \ldots}{\varepsilon^2}}$$
$$= e^{-\lambda t}\frac{1}{\varepsilon} \to \infty$$

Thus the m.s. derivative does not exist.

d) $X(t)$ is m.s. integrable if the following integral exists:

6.5. Time Average of Random Processes and Ergodic Theorems

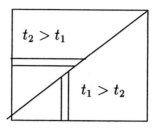

Two regions of integration

$$\int_0^t \int_0^t e^{-\lambda \max(t_1,t_2)} dt_1 dt_2$$

$$= \int_0^t dt_1 \int_0^{t_1} dt_2 e^{-\lambda t_1} + \int_0^t dt_2 \int_0^{t_2} dt_1 e^{-\lambda t_2}$$

$$= \int_0^t dt_1 t_1 e^{-\lambda t_1} + \int_0^{t_2} dt_2 t_2 e^{-\lambda t_2}$$

$$= \frac{e^{-\lambda t_1}(-\lambda t_1 - 1)}{\lambda^2}\bigg|_0^t + \frac{e^{-\lambda t_2}(-\lambda t_2 - 1)}{\lambda^2}\bigg|_0^t$$

$$= \frac{e^{-\lambda t}(-\lambda t - 1) - (-1)}{\lambda^2} + \frac{e^{-\lambda t}(-\lambda t - 1) - (-1)}{\lambda^2}$$

$$= \frac{2}{\lambda^2}[1 - e^{-\lambda t}(\lambda t + 1)]$$

$\Rightarrow$ $X(t)$ is m.s. integrable

Let
$$Y(t) = \int_0^t X(\lambda) d\lambda \ .$$

Then from Eqn. 6.84:

$$m_Y(t) = \int_0^t m_X(u) du$$
$$m_X(t) = E[X(t)] = 1 \cdot P[S < t] = 1 - e^{-\lambda t}$$
$$m_Y(t) = \int_0^t (1 - e^{-\lambda u}) du = t - \left(\frac{e^{-\lambda u}}{-\lambda}\right)_0^t$$
$$= t + \frac{1}{\lambda}[e^{-\lambda t} - 1]$$

From Eqn. 6.85 we have:

$$R_Y(t_1, t_2) = \int_0^{t_1} \int_0^{t_2} R_X(u, v) du dv$$

$$= \frac{1-e^{-\lambda t_1}(\lambda t_1+1)}{\lambda^2} + \frac{1-e^{-\lambda t_2}(\lambda t_2+1)}{\lambda^2}$$

6.79 $R_X(\tau) = A(1-|\tau|), \quad |\tau| \leq 1$

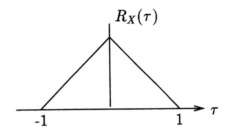

$$\begin{aligned}
VAR[<X(t)>_T] &= \frac{1}{2T}\int_{-2T}^{2T}\left(1-\frac{|u|}{2T}\right)R_X(u)du \\
&< \frac{1}{2T}\int_{-2T}^{2T} R_X(u)du \\
&= \frac{1}{2T}\int_{-2T}^{2T} A(1-|u|)du \\
&= \frac{1}{2T}\left(\frac{A}{2}\right) \quad \text{for } T>1 \\
&\to 0 \quad \text{as } T \to \infty
\end{aligned}$$

$\Rightarrow X(t)$ is mean-ergodic.

6.86 In order for $<X(t)X(t+\tau)>_T$ to be a valid estimate for $R_X(\tau)$, $Y(t) = X(t)X(t+\tau)$ must be mean-ergodic.

Note that
$$\mathcal{E}[<X(t)X(t+\tau)>_T] = \frac{1}{2T}\int_{-T}^{T}\mathcal{E}[X(t)X(t+\tau)]dt = R_X(\tau)$$
does not depend on t. Thus $X(t)X(t+\tau)$ is mean ergodic iff $C_{X(t)X(t+\tau)}(t_1,t_2)$ is such that
$$\begin{aligned}
&\lim_{T\to\infty}\frac{1}{4T^2}\int_{-T}^{T}\int_{-T}^{T}C_{X(t)X(t+\tau)}(t_1,t_2)dt_1 dt_2 \\
&= \lim_{T\to\infty}\frac{1}{4T^2}\int_{-T}^{T}\int_{-T}^{T}[\mathcal{E}[X(t_1)X(t_1+\tau)X(t_2)X(t_2+\tau)] - R_X^2(\tau)]dt_1 dt_2 \\
&= 0
\end{aligned}$$

6.5. Time Average of Random Processes and Ergodic Theorems

6.90 a) The correlation between Fourier coefficients is:

$$E[X_k X_m^*] = E\left[\frac{1}{T}\int_0^T X(t')e^{-j2\pi kt'/T}dt' \frac{1}{T}\int_0^T X(t'')e^{j2\pi mt''/T}dt''\right]$$

$$= \frac{1}{T^2}\int_0^T \int_0^T R_X(t'-t'')e^{-j2\pi kt'/T}e^{j2\pi mt''/T}dt' dt''$$

This is Eqn. 6.107.

b) Now suppose $X(t)$ is m.s. periodic:

$$E[X_k X_m^*] = \frac{1}{T^2}\int_0^T e^{j2\pi nt''/T}dt'' \int_0^T R_X(t'-t'')e^{-j2\pi k't/T}dt'$$

$$= \frac{1}{T^2}\int_0^T e^{j2\pi mt''/T}dt'' \int_{-t''}^{T-t''} R_X(u)e^{-j2\pi k(u+t'')}du$$

$$= \frac{1}{T}\int_0^T e^{j2\pi(m-k)t''/T}dt'' \frac{1}{T}\int_{-t''}^{T-t''} R_X(u)e^{-j2\pi ku}du$$

If $X(t)$ is m.s. periodic then $R_X(u)$ is periodic and the inner integral is a_k, thus

$$E[X_k X_m^*] = a_k \frac{1}{T}\int_0^T e^{j2\pi(m-k)t''/T}dt''$$

$$= a_k \delta_{km} \quad \checkmark$$

6.95 The increment of $X(t)$ in the interval $(t_1, t_2]$ has pdf:

$$f_{X(t_2)-X(t_1)}(x) = \frac{\lambda^{t_2-t_1}}{\Gamma(t_2-t_1)}x^{t_2-t_1-1}e^{-\lambda x}$$

a) We assume that $X(0) = 0$, then

$$f_{X(t_1)X(t_2)}(x,y) = f_{X(t_1)}(x)f_{X(t_2)-X(t_1)}(y-x) \quad \text{by indep. increment property}$$

$$= \frac{\lambda^{t_1}}{\Gamma(t_1)}x^{t_1}e^{-\lambda x}\frac{\lambda^{t_2-t_1}}{\Gamma(t_2-t_1)}(y-x)^{t_2-t_1-1}e^{-\lambda(y-x)}$$

$$= \frac{\lambda^{t_2}}{\Gamma(t_1)\Gamma(t_2-t_1)}x^{t_1}(y-x)^{t_2-t_1-1}e^{-\lambda y}$$

b)
$$R_X(t_1,t_2) = E[X(t_1)X(t_2)] \quad \text{assume } t_2 \geq t_1$$
$$= E[X(t_1)(X(t_2) - X(t_1) + X(t_1))]$$
$$= E[X(t_1)^2] + E[X(t_1)]E\underbrace{[X(t_2) - X(t_1)]}_{\text{increment}}$$

From Table 3.2

$$E[X(t_1)] = \frac{\alpha}{\lambda} = \frac{t_1}{\lambda}$$
$$E[X^2(t_1)] = VAR[X(t_1)] + E[X(t_1)]^2$$
$$= \frac{t_1}{\lambda^2} + \frac{t_1^2}{\lambda^2}$$
$$R_X(t_1,t_2) = \frac{t_1}{\lambda^2} + \frac{t_1^2}{\lambda^2} + \frac{t_1}{\lambda}\left(\frac{t_2 - t_1}{\lambda}\right) = \frac{t_1}{\lambda^2} + \frac{t_1 t_2}{\lambda^2}$$
$$= \frac{t_1(1 + t_2)}{\lambda^2} \quad t_2 \geq t_1$$

If $t_1 \leq t_2$, then
$$R_X(t_1,t_2) = \frac{t_2(1 + t_1)}{\lambda^2}$$

Note the similarities to the Wiener Process discussed in Ex. 6.38.

c) $R_X(t_1,t_2)$ is continuous at the point $t_2 = t_2 = t$ so $X(t)$ is m.s. continuous.

d)
$$R_X(t_1,t_2) = \begin{cases} \dfrac{t_2(1+t_1)}{\lambda^2} & t_1 < t_2 \\ \dfrac{t_2(1+t_2)}{\lambda^2} & g_2 \geq t_1 \end{cases}$$

$$\frac{\partial R_X(t_1,t_2)}{\partial t_2} = \begin{cases} \dfrac{1+t_1}{\lambda^2} & t_1 \leq t_2 \\ \dfrac{t_1}{\lambda^2} & t_2 \geq t_1 \end{cases} \Rightarrow X(t) \text{ is \underline{NOT}}$$
ms differentiable
$$= \frac{t_1}{\lambda^2} + \frac{1}{\lambda^2}u(t_1 - t_2)$$

$$R_{X'}(t_1,t_2) = \frac{\partial^2 R_X(t_1,t_2)}{\partial t_1 \partial t_2} = \frac{1}{\lambda^2} + \frac{1}{\lambda^2}\delta(t_1 - t_2)$$

This suggests that $X'(t)$ has this autocorrelation function if we generalize the notion of derivative of a random process.

Chapter 7

Analysis and Processing of Random Signals

7.1 Power Spectral Density

7.1 a) $S_X(f) = \mathcal{F}\left[g\left(\frac{\tau}{T}\right)\right] = AT \left(\frac{\sin \frac{\omega T}{2}}{\frac{\omega T}{2}}\right)^2$

Table in Appendix B.

b) $S_X(f) = g\left(\frac{f}{W}\right)$

$R_X(\tau) = AW \left(\frac{\sin \frac{W\tau}{2}}{\frac{W\tau}{2}}\right)^2$

7.6 a) $\begin{aligned} R_{XY}(\tau) &= \mathcal{E}[X(t+\tau)Y(t)] \\ &= \mathcal{E}[Y(t)X(t+\tau)] \\ &= R_{YX}(-\tau) \end{aligned}$

b) $\begin{aligned} S_{XY}(t) &= \mathcal{F}[R_{XY}(\tau)] = \int_{-\infty}^{\infty} R_{XY}(\tau) e^{-j2\pi f\tau} d\tau \\ &= \int_{-\infty}^{\infty} R_{YX}(-\tau) e^{-j2\pi f\tau} d\tau \\ &= \int_{-\infty}^{\infty} R_{YX}(\tau') e^{+j2\pi f\tau'} d\tau' \\ &= S_{YX}^*(f) \end{aligned}$

7.9 $\displaystyle\sum_{k=-\infty}^{\infty} \alpha^{|k|} e^{-j2\pi fk} = 1 + \sum_{k=1}^{\infty} \alpha^k e^{-j2\pi fk} + \sum_{k=-\infty}^{-1} \left(\frac{1}{\alpha}\right)^k e^{-j2\pi fk}$

$\displaystyle = 1 + \frac{\alpha e^{-j2\pi f}}{1 - \alpha e^{-j2\pi f}} + \frac{\alpha e^{j2\pi f}}{1 - \alpha e^{j2\pi f}}$

$\displaystyle = \frac{1 - \alpha^2}{1 + \alpha^2 - 2\alpha \cos 2\pi f}$

$\displaystyle S_X(f) = \mathcal{F}\left[4\left(\frac{1}{2}\right)^{|k|} + 16\left(\frac{1}{4}\right)^{|k|}\right]$

$\displaystyle = 4\frac{1 - \left(\frac{1}{4}\right)^2}{1 + \left(\frac{1}{2}\right)^2 - 2\left(\frac{1}{2}\right)\cos 2\pi f} + 16\frac{1 - \left(\frac{1}{4}\right)^2}{1 + \left(\frac{1}{4}\right)^2 - 2\left(\frac{1}{4}\right)\cos 2\pi f}$

$\displaystyle = \frac{12}{5 - 4\cos 2\pi f} + \frac{240}{17 - 8\cos 2\pi f}$

7.12 a) $\mathcal{E}[D_n] = \mathcal{E}[X_n] - \mathcal{E}[X_{n-d}] = 0$

$\begin{aligned} R_D(n, n+k) &= \mathcal{E}[(X_n - X_{n-d})(X_{n+k} - X_{n+k-d})] \\ &= R_X(k) - R_X(d+k) - R_X(k-d) + R_X(k) \\ &= 2R_X(k) - R_X(k+d) - R_X(k-d) \\ S_D(f) &= 2S_X(f) - S_X(f)e^{j2\pi fd} - S_X(f)e^{-j2\pi fd} \\ &= 2S_X(f)(1 - \cos 2\pi fd) \end{aligned}$

b) $\mathcal{E}[D_n^2] = R_D(0) = 2R_X(0) - 2R_X(d)$

7.2 Response of Linear Systems to Random Signals

7.17 a) $S_Y(f) = |H(f)|^2 S_X(f) = 4\pi^2 f^2 S_X(f)$

b) $R_Y(\tau) = \mathcal{F}^{-1}[S_Y(f)] = -\frac{d^2}{d\tau^2} R_X(\tau)$

7.2. Response of Linear Systems to Random Signals

7.21 a) $S_{YX}(f) = H(f)S_X(f) = \dfrac{N_0/2}{1+j2\pi f}$

$R_{YX} = \mathcal{F}^{-1}[S_{YX}(f)] = \dfrac{N_0}{2}e^{-\tau} \quad \tau > 0$

b) $S_Y(f) = |H(f)|^2 S_X(f) = \dfrac{N_0/2}{1+4\pi^2 f^2}$

$R_Y(\tau) = \mathcal{F}^{-1}[S_Y(f)] = \dfrac{N_0}{4}e^{-|\tau|}$

c) $R_Y(0) = \dfrac{N_0}{4}$

7.26

$$
\begin{aligned}
R_Z(\tau) &= \mathcal{E}[Z(t+\tau)Z(t)] = \mathcal{E}[(X(t+\tau)-Y(t+\tau))(X(t)-Y(t))] \\
&= R_X(\tau) + R_Y(\tau) - R_{XY}(\tau) - R_{YX}(\tau) \\
R_{XY}(\tau) &= \mathcal{E}[X(t+\tau)Y(t)] = \mathcal{E}\left[X(t+\tau)\int_{-\infty}^{\infty} h(\lambda)V(t-\lambda)d\lambda\right] \\
&= \int_{-\infty}^{\infty} h(\lambda) R_{XV}(\tau+\lambda)d\lambda \\
&= \int_{-\infty}^{\infty} h(\lambda) R_X(\tau+\lambda)d\lambda \quad \text{since} \quad \begin{aligned} R_{XV}(\tau) &= \mathcal{E}[X(t+\tau)(X(t)+N(t))] \\ &= R_X(\tau)\end{aligned} \\
&= h(-\tau) \star R_X(\tau) \\
S_Z(f) &= S_X(f) + S_Y(f) - S_{XY}(f) - S_{YX}(f) \\
&= S_X(f) + |H(f)|^2(S_X(f)+S_N(f)) - H(f)S_X(f) - H^*(f)S_X(f) \\
S_Z(f) &= |1-H(f)|^2 S_X(f) + |H(f)|^2 S_N(f) \quad\quad (*)
\end{aligned}
$$

Comments: If we view $Y(t)$ as our estimate for $X(t)$, then $S_Z(f)$ is the power spectral density of the error signal $Z(t) = Y(t) - X(t)$. Equation (*) suggests the following:

if $S_X(f) \gg S_N(f)$ let $H(f) \approx 1$
if $S_X(f) \ll S_N(f)$ let $H(f) \approx 0$

i.e. select $H(f)$ to "pass" the signal and reject the noise.

7.29 a) $H(f) = \sum_{k=0}^{\infty} \left(\frac{1}{2}\right)^n e^{-j2\pi f n} = \frac{1}{1 - \frac{1}{2}e^{-j2\pi f}}$

$G(f) = \sum_{k=0}^{\infty} \left(\frac{1}{4}\right)^n e^{-j2\pi f n} = \frac{1}{1 - \frac{1}{4}e^{-j2\pi f}}$

$S_Y(f) = |H(f)|^2 \frac{N_0}{2} = \frac{N_0/2}{\frac{5}{4} - \cos 2\pi f}$

$S_Z(f) = |G(f)|^2 S_Y(f) = \frac{N_0/2}{\left(\frac{5}{4} - \cos 2\pi f\right)\left(\frac{17}{16} - \frac{1}{2}\cos 2\pi f\right)}$

b) $S_{WY}(f) = H(f)S_X(f) = \frac{N_0/2}{1 - \frac{1}{2}e^{-j2\pi f}}$

$\Rightarrow R_{WY}(k) = \frac{N_0}{2}\left(\frac{1}{2}\right)^k u(k)$

$S_{WZ}(f) = H(f)G(f)S_X(f) = \frac{N_0/2}{(1 - \frac{1}{2}e^{-j2\pi f})(1 - \frac{1}{4}e^{-j2\pi f})}$

$= \frac{N_0}{1 - \frac{1}{2}e^{-j2\pi f}} - \frac{N_0/2}{1 - \frac{1}{4}e^{-j2\pi f}}$

$R_{WZ}(k) = \left(N_0\left(\frac{1}{2}\right)^k - \frac{N_0}{2}\left(\frac{1}{4}\right)^k\right)u(k)$

c) $S_Z(f) = \frac{N_0/2}{\left(\frac{5}{4} - \cos 2\pi f\right)\left(\frac{17}{16} - \frac{1}{2}\cos 2\pi f\right)} = \frac{\frac{8}{7}N_0}{\frac{5}{4} - \cos 2\pi f} - \frac{\frac{4}{7}N_0}{\frac{17}{16} - \frac{1}{2}\cos 2\pi f}$

From Problem 7.9 we know that

$\mathcal{F}[|\alpha|^k] = \sum_{k=-\infty}^{\infty} |\alpha|^k e^{-j2\pi f k} = \frac{1 - \alpha^2}{1 + \alpha^2 - 2\alpha\cos 2\pi f}$

$\therefore R_Z(k) = \frac{8}{7}N_0 \frac{4}{3}\mathcal{F}^{-1}\left[\frac{1 - \left(\frac{1}{2}\right)^2}{1 + \left(\frac{1}{2}\right)^2 - 2\left(\frac{1}{2}\right)\cos 2\pi f}\right]$

7.2. Response of Linear Systems to Random Signals

$$-\frac{4}{7}N_0\frac{16}{15}\mathcal{F}^{-1}\left[\frac{1-\left(\frac{1}{4}\right)^2}{1+\left(\frac{1}{4}\right)^2-2\left(\frac{1}{4}\right)\cos 2\pi f}\right]$$

$$=\frac{32}{21}N_0\left(\frac{1}{2}\right)^{|k|}-\frac{64}{105}N_0\left(\frac{1}{4}\right)^{|k|}$$

$$\mathcal{E}[Z_n^2] = R_Z(0)$$
$$= \left(\frac{32}{21}-\frac{64}{105}\right)N_0$$
$$= \frac{96}{105}N_0$$

7.36 a)
$$\mathcal{E}[Y_n^2] = \mathcal{E}\left[Y_n\left(\sum_{i=1}^{q}\alpha_i Y_{n-i}+W_n\right)\right]$$
$$= \sum_{i=1}^{q}\alpha_i R_Y(i) + R_{YW}(0)$$

$$R_{YW}(0) = \mathcal{E}\left[\left(\sum_{i=1}^{q}\alpha_i Y_{n-i}+W_n\right)W_n\right]$$
$$= \sum_{i=1}^{q}\alpha_i \underbrace{\mathcal{E}[Y_{n-i}W_n]}_{0}+R_W(0) = R_W(0)$$

$$\therefore R_Y(0) = \sum_{i=1}^{q}\alpha_i R_Y(i) + R_W(0)$$

$$R_Y(k) = \mathcal{E}\left[Y_{n-k}\left(\sum_{i=1}^{q}\alpha_i Y_{n-i}+W_n\right)\right]$$
$$= \sum_{i=1}^{q}\alpha_i R_Y(k-i) + \underbrace{\mathcal{E}[Y_{n-k}W_n]}_{\underbrace{\mathcal{E}[Y_{n-k}]\mathcal{E}[W_n]}_{0}}$$
$$= \sum_{i=1}^{q}\alpha_i R_Y(k-i)$$

b)
$$Y_n = rY_{n-1}+W_n$$
$$R_Y(0) = rR_Y(1)+R_W(0)$$
$$R_Y(k) = rR_Y(k-1) \Rightarrow R_Y(1) = rR_Y(0)$$
$$\Rightarrow R_Y(0) = r^2 R_Y(0)+R_W(0) \Rightarrow R_Y(0) = \frac{R_W(0)}{1-r^2}$$

$$\Rightarrow R_Y(k) = \begin{cases} \frac{r^k R_W(0)}{1-r^2} = \underbrace{\left(\frac{R_W(0)}{1-r^2}\right)}_{\sigma_Y^2} r^k & k > 0 \\ R_Y(-k) = \sigma_Y^2 r^{-|k|} & k < 0 \end{cases}$$

7.3 Amplitude Modulation by Random Signals

7.37 $Y(t) = \underbrace{A(t)\cos(2\pi f_c t + \Theta)}_{X(t)} + N(t)$

Assuming $X(t)$ and $N(t)$ are independent random processes:
$$R_Y(\tau) = R_X(\tau) + R_N(\tau)$$

From Example 7.4 and the fact that $\mathcal{E}[X(t)] = 0$
$$\begin{aligned} S_Y(t) &= S_X(\tau) + S_N(f) \\ &= \frac{1}{2}S_A(f - f_c) + \frac{1}{2}S_A(f + f_c) + S_N(f) \end{aligned}$$

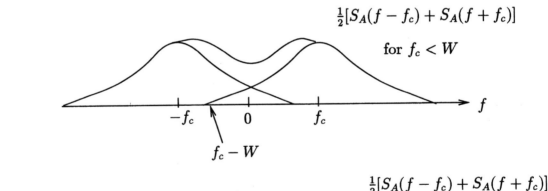

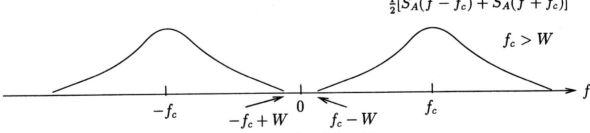

where we assumed that $S_A(f)$ is bandlimited to $|f| < W$.

7.4 Optimum Linear Filters

7.44 $X_\alpha = Z_\alpha + N_\alpha$ first order AR, $R_Z(k) = \sigma_Z^2 r^{|k|}$

$$\begin{bmatrix} 1+\Gamma & r \\ r & 1+\Gamma \end{bmatrix} \begin{bmatrix} k_0 \\ k_1 \end{bmatrix} = \begin{bmatrix} 1 \\ r \end{bmatrix}$$

where $\Gamma = \dfrac{\sigma_N^2}{\sigma_Z^2}$.

$$\begin{bmatrix} k_0 \\ k_1 \end{bmatrix} = \frac{1}{(1+\Gamma)^2 - r^2} \begin{bmatrix} 1+\Gamma & -r \\ -r & 1+\Gamma \end{bmatrix} \begin{bmatrix} 1 \\ r \end{bmatrix} = \frac{1}{(1+\gamma)^2 - r^2} \begin{bmatrix} 1+\Gamma - r^2 \\ \Gamma r \end{bmatrix}$$

$$\begin{aligned}
\mathcal{E}[(Z_t - Y_t)^2] &= R_Z(0) - \sum_{\beta=0}^{1} k_\beta R_{ZX}(\beta) = R_Z(0) - \sum_{\beta=0}^{1} k_\beta \sigma_Z^2 r^{|\beta|} \\
&= \sigma_Z^2 \left[1 - \frac{1-\Gamma - r^2}{(1+\Gamma)^2 - r^2} - \frac{\Gamma r}{(1+\Gamma)^2 - r^2} r \right] \\
&= \sigma_Z^2 \left[1 - \frac{(1+\Gamma)(1-r^2)}{(1+\Gamma)^2 - r^2} \right] = \frac{4}{7} \sigma_Z^2
\end{aligned}$$

since $\Gamma = \frac{1}{4}$, $r = \frac{1}{2}$.

7.60 $X_n = Z_n + N_n \quad R_Z(k) = 4 \left(\frac{1}{2}\right)^{|k|} \quad R_N(k) = \delta_k$

$$\begin{aligned}
S_X(f) &= S_Z(f) + S_N(f) = \frac{4}{\frac{5}{4} - \cos 2\pi f} + 1 = \frac{\frac{21}{4} - \cos 2\pi f}{\frac{5}{4} - \cos 2\pi f} \\
&= \frac{\frac{Z_2}{2}(1 - Z_2 e^{-j2\pi f})(1 - Z_1 e^{+j2\pi f})}{(1 - \frac{1}{2}e^{-j2\pi f})(1 - \frac{1}{2}e^{j2\pi f})}
\end{aligned}$$

after factoring the numerator and denominator

where

$$Z_1 = \frac{\frac{21}{4} - \sqrt{\left(\frac{21}{4}\right)^2 - 4}}{2} = \frac{1}{5}$$

$$Z_2 = \frac{\frac{21}{4} + \sqrt{\left(\frac{21}{4}\right)^2 - 4}}{2} = \frac{1}{Z_1}$$

$$\Rightarrow G(f) = \sqrt{\frac{Z_2}{2}} \frac{1 - Z_1 e^{-j2\pi f}}{1 - \frac{1}{2}e^{-j2\pi f}} = \frac{1}{W(f)}$$

Next consider:

$$R_{ZX}(k) = \mathcal{E}[Z_{n+n}(Z_n + N_n)] = R_Z(k)$$
$$\Rightarrow S_{ZX}(f) = S_Z(f)$$

Thus

$$\begin{aligned}
S_{ZX'}(f) &= W^*(f) S_{ZX}(f) \\
&= \sqrt{\frac{2}{Z_2}} \frac{1 - \frac{1}{2}e^{+j2\pi f}}{1 - Z_1 e^{+j2\pi f}} S_Z(f) \\
&= \sqrt{\frac{2}{Z_2}} \left(\frac{1}{1 - Z_1 e^{+j2\pi f}}\right) \left(\frac{4}{1 - \frac{1}{2}e^{-j2\pi f}}\right) \\
&= \frac{4\sqrt{\frac{2}{Z_2}}}{1 - \frac{1}{2}Z_1} \left[\frac{1}{1 - Z_1 e^{j2\pi f}} + \underbrace{\frac{1}{1 - \frac{1}{2}e^{-j2\pi f}}}_{\text{this yields the positive time component}}\right] \quad \text{after partial fraction expansion}
\end{aligned}$$

$$\therefore H_2(f) = \frac{4\sqrt{\frac{2}{Z_2}}}{1 - \frac{1}{2}Z_1} \frac{1}{1 - \frac{1}{2}e^{-j2\pi f}}$$

and finally

$$\begin{aligned}
H(f) &= W(f) H_2(f) = \sqrt{\frac{2}{Z_2}} \frac{1 - \frac{1}{2}e^{-j2\pi f}}{1 - Z_1 e^{-j2\pi f}} \left(\frac{4\sqrt{\frac{2}{Z_2}}}{1 - \frac{1}{2}Z_1}\right) \left(\frac{1}{1 - \frac{1}{2}e^{-j2\pi f}}\right) \\
&= \frac{\frac{8}{Z_2(1 - \frac{1}{2}Z_1)}}{1 - Z_1 e^{-j2\pi f}} \\
&= \left(\frac{8}{Z_2 - \frac{1}{2}}\right) \frac{1}{1 - Z_1 e^{-j2\pi f}}
\end{aligned}$$

7.5. A Computer Method for Estimating the Power Spectral Density

$$h_n = \left(\frac{8}{Z_2 - \frac{1}{2}}\right) Z_1^k \quad k \geq 0$$

is the impulse response of the optimum filter.

7.5 A Computer Method for Estimating the Power Spectral Density

7.67
$$\tilde{p}_k(f) = \frac{1}{k}\left|\sum_{l=0}^{k-1} X_l e^{-j2\pi fl}\right|^2$$

$$= \frac{1}{k}\sum_{l=0}^{k-1}\sum_{l'=0}^{k-1} X_l X_{l'} e^{-j2\pi fl} e^{+j2\pi fl'}$$

$$= \frac{1}{k}\sum_{l=0}^{k-1}\sum_{l'=0}^{k-1} X_l X_{l'} e^{-j2\pi f(l-l')}$$

$$m = l - l' \Leftrightarrow l' = l - m$$

where $(k-1) \leq m \leq k-1$

We change the order of summation as indicated by the figure below:

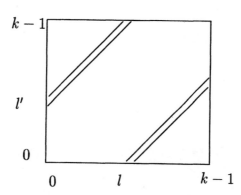

We then obtain:

$$= \sum_{m=-(k-1)}^{k-1}\left\{\underbrace{\frac{1}{k}\sum_{l=0}^{k-|m|-1} X_l X_{l+n}}_{\hat{r}_k(m)}\right\} e^{-j2\pi fm}$$

$$= \sum_{m=-(k-1)}^{k-1} \hat{r}_k(m) e^{-j2\pi fm} \quad \checkmark$$

7.70

$$\mathcal{E}\left[\sum_{m=-(k-1)}^{k-1} \left(\frac{1}{k-|m|} \sum_{n=0}^{k-|m|-1} X_n X_{n+m} \right) e^{-j2\pi fm} \right]$$

$$= \sum_{m=(k-1)}^{k-1} \frac{1}{k-|m|} \left(\underbrace{\sum_{n=0}^{k-|m|-1} R_X(m)}_{(k-|m|)R_X(m)} \right) e^{-j2\pi fm}$$

$$= \sum_{m=-(k-1)}^{k-1} R_X(m) e^{-j2\pi fm}$$

The estimate is biased because the limits of the summation are finite.

Chapter 8

Markov Chains

8.1 Markov Processes

8.2 a) The number X_n of black balls in the urn completely specifies the probability of outcomes of a trial; therefore X_n is independent of its past values and X_n is a Markov process.

$$P[X_n = 4|X_{n-1} = 5] = \frac{5}{10} = 1 - P[X_n = 5|X_{n-1} = 5]$$
$$P[X_n = 3|X_{n-1} = 4] = \frac{4}{9} = 1 - P[X_n = 4|X_{n-1} = 4]$$
$$P[X_n = 2|X_{n-1} = 3] = \frac{3}{8} = 1 - P[X_n = 3|X_{n-1} = 3]$$
$$P[X_n = 1|X_{n-1} = 2] = \frac{2}{7} = 1 - P[X_n = 2|X_{n-1} = 2]$$
$$P[X_n = 0|X_{n-1} = 1] = \frac{1}{6} = 1 - P[X_n = 1|X_{n-1} = 1]$$
$$P[X_n = 0|X_{n-1} = 0] = 1$$

All transition prob. are independent of time.

8.2 Discrete-Time Markov Chains

8.8   a) $P = \begin{bmatrix} 1 & 0 & 0 & 0 & 0 & 0 \\ \frac{1}{6} & \frac{5}{6} & 0 & 0 & 0 & 0 \\ 0 & \frac{2}{7} & \frac{5}{7} & 0 & 0 & 0 \\ 0 & 0 & \frac{3}{8} & \frac{5}{7} & 0 & 0 \\ 0 & 0 & 0 & \frac{4}{9} & \frac{5}{9} & 0 \\ 0 & 0 & 0 & 0 & \frac{5}{10} & \frac{5}{10} \end{bmatrix}$

b) $P^2 = \begin{bmatrix} 1 & 0 & 0 & 0 & 0 & 0 \\ \frac{11}{36} & \frac{25}{36} & 0 & 0 & 0 & 0 \\ \frac{1}{21} & \frac{65}{144} & \frac{25}{49} & 0 & 0 & 0 \\ 0 & \frac{3}{28} & \frac{225}{448} & \frac{25}{64} & 0 & 0 \\ 0 & 0 & \frac{1}{6} & \frac{95}{192} & \frac{25}{81} & 0 \\ 0 & 0 & 0 & \frac{2}{9} & \frac{19}{36} & \frac{1}{4} \end{bmatrix}$

$$p_{54}(2) = \frac{19}{36} \quad \text{from } P^2$$

$$p_{54}(2) = p_{55}^{\text{no change}}(1)p_{54}^{\text{change}}(1) + p_{54}^{\text{change}}(1)p_{44}^{\text{no change}}(1)$$

$$= \frac{1}{2}\frac{1}{2} + \frac{1}{2}\frac{5}{9} = \frac{19}{36} \quad \checkmark$$

c) As $n \to \infty$ eventually all black balls are removed. Thus

$$P^n \to \begin{bmatrix} 1 & 0 & 0 & 0 & 0 & 0 \\ 1 & 0 & 0 & 0 & 0 & 0 \\ 1 & 0 & 0 & 0 & 0 & 0 \\ 1 & 0 & 0 & 0 & 0 & 0 \\ 1 & 0 & 0 & 0 & 0 & 0 \\ 1 & 0 & 0 & 0 & 0 & 0 \end{bmatrix}$$

8.12 $X_n \epsilon \{0, 1\}$ where $0 = $ working, $1 = $ not working

a) $P = \begin{bmatrix} 1-a & a \\ b & 1-b \end{bmatrix}$

8.2. Discrete-Time Markov Chains

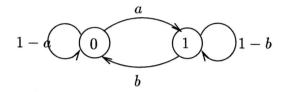

b) To find the eigenvalues, consider

$$|P - \lambda I| = (1 - b - \lambda)(1 - a - \lambda) - ab = 0$$
$$\Rightarrow \lambda_1 = 1 \quad \lambda_2 = 1 - a - b$$

Then the eigenvectors are $\underline{e}_1 = [1, \frac{b}{a}]$, $\underline{e}_2 = [1, -1]$, so

$$E = \begin{bmatrix} \underline{e}_1 \\ \underline{e}_2 \end{bmatrix} = \begin{bmatrix} 1 & \frac{b}{a} \\ 1 & -1 \end{bmatrix}$$

and

$$E^{-1} = \frac{1}{a+b} \begin{bmatrix} a & b \\ a & -a \end{bmatrix}$$

and thus

$$\begin{aligned} P^n &= E^{-1} \begin{bmatrix} 1 & 0 \\ 0 & (1-a-b)^n \end{bmatrix} E \\ &= \frac{1}{a+b} \begin{bmatrix} a + b(1-a-b)^n & b - b(1-a-b)^n \\ a - a(1-a-b)^n & b + c(1-a-b)^n \end{bmatrix} \end{aligned}$$

c) $0 < a + b < 2$ since $0 < a < 1$ and $0 < b < 1$
$\Rightarrow -1 < 1 - a - b < 1 \Rightarrow (1-a-b)^n \to 0$

$$\therefore P^n \to \begin{bmatrix} \frac{a}{a+b} & \frac{b}{a+b} \\ \frac{a}{a+b} & \frac{b}{a+b} \end{bmatrix}$$

and

$$\underline{p}(n) \to [\frac{a}{a+b}, \frac{b}{a+b}]$$

8.3 Continuous-Time Markov Chains

8.14 From Ex. 8.15 we have

$$p_0(t) = \frac{\beta}{\alpha+\beta} + \left(p_0(0) - \frac{\beta}{\alpha+\beta}\right) e^{-(\alpha+\beta)t}$$

$$p_1(t) = \frac{\alpha}{\alpha+\beta} + \left(p_1(0) - \frac{\alpha}{\alpha+\beta}\right) e^{-(\alpha+\beta)t}$$

a) Now suppose we know the initial state is 0, then $p_0(0) = 1 \Rightarrow$

$$p_{00}(t) = \frac{\beta}{\alpha+\beta} + \left(1 - \frac{\beta}{\alpha+\beta}\right) e^{-(\alpha+\beta)t} = \frac{\beta + \alpha e^{-(\alpha+\beta)t}}{\alpha+\beta}$$

$$p_{01}(t) = 1 - p_{00}(t) = \frac{\alpha(1 - e^{-(\alpha+\beta)t})}{\alpha+\beta}$$

If the initial state is 1, then $p_1(0) = 1 \Rightarrow$

$$p_{11}(t) = \frac{\alpha}{\alpha+\beta} + \left(1 + \frac{\alpha}{\alpha+\beta}\right) e^{-(\alpha+\beta)t} = \frac{\alpha + \beta e^{-(\alpha+\beta)t}}{\alpha+\beta}$$

$$p_{10}(t) == 1 - p_{11}(t) = \frac{\beta(1 - e^{-(\alpha+\beta)t}}{\alpha+\beta}$$

$$\therefore \mathbf{P}(t) = \frac{1}{\alpha+\beta} \begin{bmatrix} \beta + \alpha e^{-(\alpha+\beta)t} & \alpha(1 - e^{-(\alpha+\beta)t}) \\ \beta(q - e^{-(\alpha+\beta)t}) & \alpha + \beta e^{-(\alpha+\beta)t} \end{bmatrix}$$

b) $P[X(1.5) = 1, X(3) = 1/X(0) = 0]$

$$= P[X(3) = 1/X(1.5) = 1, X(0) = 0] P[X(1.5) = 1/X(0) = 0]$$
$$= P[X(3) = 1/X(1.5) = 1] P[X(1.5) = 1/X(0) = 0]$$
$$= p_{11}(1.5) p_{01}(1.5)$$

$$P[X(1.5) = 1, X(3) = 1] = P[X(3) = 1/X(1.5) = 1] P[X(1.5) = 1]$$
$$= p_{11}(1.5) \left[\frac{\alpha}{\alpha+\beta} + \left(p_1(0) - \frac{\alpha}{\alpha+\beta}\right) e^{-(\alpha+\beta)1.5}\right]$$

8.3. Continuous-Time Markov Chains

8.19 The transition rate diagram is:

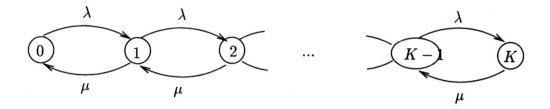

Eqn. 8.33 applies here, so we have

$$P_{j+1} = \left(\frac{\lambda}{\mu}\right) P_j = \left(\frac{\lambda}{\mu}\right)^{j+1} P_0$$

To find P_0 consider

$$1 = P_0 \sum_{j=0}^{K} \left(\frac{\lambda}{\mu}\right)^j = P_0 \frac{1 - \left(\frac{\lambda}{\mu}\right)^{K+1}}{1 - \frac{\lambda}{\mu}}$$

$$\Rightarrow P_j = \frac{\left(1 - \frac{\lambda}{\mu}\right)}{1 - \left(\frac{\lambda}{\mu}\right)^{K+1}} \left(\frac{\lambda}{\mu}\right)^j \quad 0 \leq j \leq K$$

8.4 Classes of States, Recurrence Properties, and Limiting Probabilities

8.20 a)

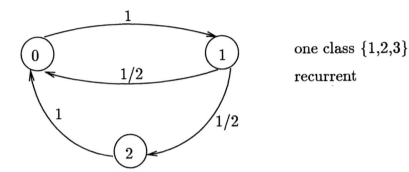

one class {1,2,3}

recurrent

8.21 a) This is a positive recurrent Markov chain.
For state 0 the mean first return time is (see solution of Prob. 8.20)

$$\mathcal{E}[T_0] = \underbrace{2 \cdot \frac{1}{2}}_{0 \to 1 \to 0} + \underbrace{3 \cdot \frac{1}{2}}_{0 \to 1 \to 2 \to 0} = \frac{5}{2} \Rightarrow \pi_0 = \frac{2}{5}$$

For state 1

$$\mathcal{E}[T_1] = \underbrace{2 \cdot \frac{1}{2}}_{1 \to 0 \to 1} + \underbrace{3 \cdot \frac{1}{2}}_{1 \to 2 \to 0 \to 1} = \frac{5}{2} \Rightarrow \pi_1 = \frac{2}{5}$$

For state 2

$$\begin{aligned}\mathcal{E}[T_2] &= 3 \cdot \frac{1}{2} + 5\left(\frac{1}{4}\right) + 7\frac{1}{8} + \ldots \\ &= \sum_{j=1}^{\infty}(2j+1)\left(\frac{1}{2}\right)^j = 2\underbrace{\sum_{j=1}^{\infty} j\left(\frac{1}{2}\right)^j}_{\frac{\frac{1}{2}}{\left(1-\left(\frac{1}{2}\right)\right)^2}}\end{aligned}$$

$$\Rightarrow \pi_2 = \frac{1}{5}$$

It can easily be shown that above are solutions to equations for the stationary proof.

8.5 Time-Reversed Markov Chains

8.26 a) The state transition diagram is:

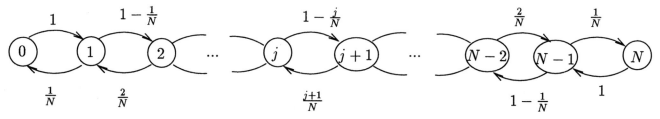

This is a birth-death process, so by Example 8.37 it is reversible. If process is time reversible then

$$\pi_i p_{ij} = \pi_j p_{ji}$$
$$\Rightarrow \pi_i p_{i,i+1} = \pi_{i+1} p_{i+1,i}$$
$$\pi_i \left(1 - \frac{1}{N}\right) = \pi_{i+1} \frac{i+1}{N}$$
$$\pi_{i+1} = \frac{N-i}{i+1} \pi_i$$

It is then easy to show that

$$\pi_j = \binom{N}{j} \pi_0$$

$$1 = \pi_0 \sum_{j=0}^{N} \binom{N}{j} = \pi_0 2^N$$

$$\Rightarrow \pi_j = \binom{N}{j} \left(\frac{1}{2}\right)^N$$

Chapter 9

Introduction to Queueing Theory

9.1 & 9.2 The Elements of a Queueing Network and Little's Formula

9.2 $\{S_i\} = \{1, 3, 4, 7, 8, 15\}$
$\{\tau_i\} = \{3.5, 4, 2, 1, 1.5, 4\}$

a) FCFS

i	S_i	τ_i	D_i	W_i	T_i
1	1	3.5	4.5	0	3.5
2	3	4	3.5	1.5	5.5
3	4	2	10.5	4.5	6.5
4	7	1	11.5	3.5	4.5
5	8	1.5	13.0	3.5	5.0
6	15	4			

where $W_i = D_{i-1} - S_i = T_i - \tau_i$ and $T_i = D_i - S_i = W_i + \tau_i$

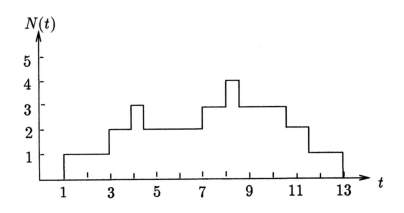

$$<N>_{13} = \frac{1}{13}\sum_{i=1}^{A_{13}} T_i = \frac{25}{13}$$

$$<\lambda>_{13} = \frac{A_{13}}{13} = \frac{5}{13}$$

$$<T>_{13} = \frac{1}{A_{13}}\sum_{i=1}^{A_{13}} T_i = \frac{25}{5}$$

$$<N>_{13} = \frac{25}{13} = <\lambda>_{13}<T>_{13} = \frac{5}{13}\frac{25}{5} \quad \checkmark$$

b) LCFS

i	S_i	τ_i	D_i	$W_i = T_i - \tau_i$	$T_i = D_i - S_i$
1	1	3.5	4.5	0	3.5
2	3	4	10.5	3.5	7.5
3	4	2	6.5	0.5	2.5
4	7	1	13.0	5.0	6.0
5	8	1.5	12.0	2.5	4.0

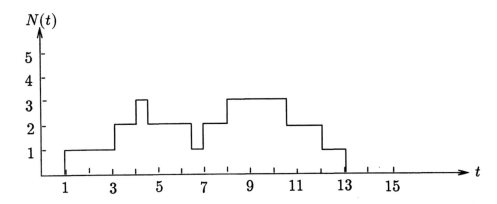

9.1 & 9.2 The Elements of a Queueing Network and Little's Formula

$$<N>_{13} = \frac{23.5}{13} \quad <\lambda>_{13} = \frac{5}{13} \quad <T>_{13} = \frac{23.5}{5}$$
$$<N>_{13} = <\lambda>_{13}<T>_{13}$$

c) Shortest Job First:

i	S_i	τ_i	D_i	$W_i = T_i - \tau_i$	$T_i = D_i - S_i$
1	1	3.5	4.5	0	3.5
2	3	4	10.5	3.5	7.5
3	4	2	6.5	0.5	2.5
4	7	1	11.5	3.5	4.5
5	8	1.5	13.0	3.5	5.0

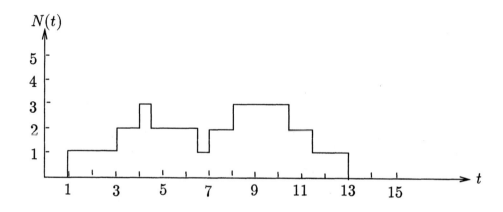

$$<N>_{13} = \frac{23}{13} \quad <\lambda>_{13} = \frac{5}{13} \quad <T>_{13} = \frac{23}{5}$$
$$<N>_{13} = <\lambda>_{13}<T>_{13}$$

9.5 a) $\lambda T = 5$

b) $\lambda m = 2$

c) $T = \dfrac{5}{\lambda} = 5\left(\dfrac{m}{2}\right) = \dfrac{5}{2}m$

9.3 The M/M/1 Queue

9.17 If $N < K$ arrival rate is λ
If $N \geq K$ arrival rate is reduced to $\frac{\lambda}{2}$

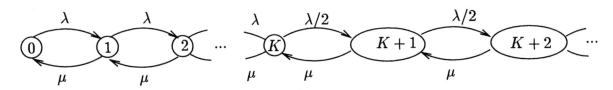

For $0 \leq j \leq K$
$$P_j = \frac{\lambda}{\mu} P_{j-1} = \left(\frac{\lambda}{\mu}\right)^j P_0$$

For $K < j$
$$P_j = \frac{\lambda}{2\mu} P_{j-1} = \left(\frac{\lambda}{2\mu}\right)^{j-K} P_K = \left(\frac{\lambda}{2\mu}\right)^{j-K} \left(\frac{\lambda}{\mu}\right)^K P_0$$

$$1 = \sum_{j=0}^{\infty} P_j = P_0 \underbrace{\sum_{j=0}^{K-1} \left(\frac{\lambda}{\mu}\right)^j}_{\frac{1-\left(\frac{\lambda}{\mu}\right)^K}{1-\frac{\lambda}{\mu}}} + P_0 \underbrace{\sum_{j=K}^{\infty} \left(\frac{\lambda}{2\mu}\right)^{j-K} \left(\frac{\lambda}{\mu}\right)^K}_{\frac{\left(\frac{\lambda}{\mu}\right)^K}{1-\frac{\lambda}{2\mu}}}$$

$$P_0 = \left[\frac{1-\left(\frac{\lambda}{\mu}\right)^K}{1-\frac{\lambda}{\mu}} + \frac{\left(\frac{\lambda}{\mu}\right)^K}{1-\frac{\lambda}{2\mu}}\right]^{-1}$$

9.4 Multi-Server Systems: M/M/c, M/M/c/c, and M/M/∞

9.18
$$P[N \geq c+k] = \sum_{j=c+k}^{\infty} \frac{\rho^{j-c}}{c!} a^c P_0 = \frac{a^c}{c!} P_0 \rho^k \sum_{j'=0}^{\infty} \rho^j$$
$$= \frac{\frac{a^c}{c!} P_0 \rho^k}{1-\rho} = \frac{p_c \rho^k}{1-\rho}$$

9.4. Multi-Server Systems: M/M/c, M/M/c/c, and M/M/∞

9.19 $\lambda = 12 \quad \frac{1}{\mu} = \frac{5}{60} \quad c = 2$
$\Rightarrow a = \frac{\lambda}{\mu} = 1 \quad \rho = \frac{a}{2} = \frac{1}{2}$

a) $\quad P[N \geq c] = \dfrac{\frac{a^c}{c!} P_0}{1 - rho} = c(c, a)$

$$P_0 = \left\{ \sum_{j=0}^{1} \frac{a^j}{j!} + \frac{a^2}{2!} \sum_{j=0}^{\infty} \rho^j \right\}^{-1} = \left\{ 1 + 1 + \frac{1}{2} \frac{1}{1 - \frac{1}{2}} \right\}^{-1} = \frac{1}{3}$$

$\Rightarrow P[N \geq 2] = \dfrac{\frac{1}{2!}\frac{1}{3}}{1 - \frac{1}{2}} = \dfrac{1}{3} = C(2, 1)$

b) $\mathcal{E}[N] = \mathcal{E}[N_q] + a = \dfrac{\rho}{1-\rho} C(c, a) + a = \dfrac{\frac{1}{2}}{1 - \frac{1}{2}} \dfrac{1}{3} + 1 = \dfrac{4}{3}$

$\mathcal{E}[T] = \dfrac{1}{\lambda} \mathcal{E}[N] = \dfrac{1}{9}$

c) $P[N > 4] = P[N_q > 2] = \sum_{j=3}^{\infty} \rho^{j-2} P_2 = \dfrac{P_2 \rho}{1-\rho} = \dfrac{1}{6}$

9.24 $\lambda = 10 \quad \dfrac{1}{\mu} = \dfrac{1}{2} \quad \dfrac{\lambda}{\mu} = 5 = a$

$$B(0, 5) = 1$$

$$B(1, 5) = \dfrac{5 \cdot 1}{1 + 5 \cdot 1} = \dfrac{5}{6}$$

$$B(2, 5) = \dfrac{5 \left(\frac{5}{6}\right)}{2 + 5 \left(\frac{5}{6}\right)} = \dfrac{25}{37}$$

$$B(3, 5) = \dfrac{5 \left(\frac{25}{37}\right)}{3 + 5 \left(\frac{25}{37}\right)} = \dfrac{125}{236}$$

$$B(4, 5) = \dfrac{5 \left(\frac{125}{236}\right)}{4 + 5 \left(\frac{125}{236}\right)} = \dfrac{625}{1569}$$

$$\rightarrow B(5, 5) = \dfrac{5 \left(\frac{625}{1569}\right)}{5 + 5 \left(\frac{625}{1569}\right)} = \dfrac{625}{2194} \approx 28.5\%$$

$$\vdots$$
$$B(8,5) = 0.070 \quad \text{need 3 more servers}$$

9.5 Finite Source Queueing Systems

9.31

$$P[N_a = k] = \frac{\frac{(K-1)!(\alpha/\mu)^k}{(K-1-k)!}}{\sum_{k'=0}^{K-1} \frac{(K-1)!(\alpha/\mu)^{k'}}{(K-1-k')!}} = \frac{\frac{(\alpha/\mu)^k}{(K-1-k)!}}{\sum_{k'=0}^{K-1} \frac{(\alpha/\mu)^{k'}}{(K-1-k')!}}$$

$$\begin{aligned}
\mathcal{E}[T] &= \frac{1}{\mu} \sum_{k=0}^{K-1} (k+1) P[N_a = k] \\
&= \frac{1}{\mu} \sum_{k=0}^{K-1} (k+1) \frac{\frac{(\alpha/\mu)^k}{(K-1-k)!}}{\sum_{k'=0}^{K-1} \frac{(\alpha/\mu)^{k'}}{(K-1-k')!}} \quad \text{Let } j = K-1-k, j' = K-1-k' \\
&= \frac{1}{\mu} \sum_{j=0}^{K-1} (K-j) \underbrace{\frac{\frac{(\mu/\alpha)^j}{j!}}{\sum_{j'=0}^{K-1} \frac{(\mu/\alpha)^{j'}}{j'!}}}_{\text{probs of M/M/K-1/K-1}} \\
&= \frac{1}{\mu} \left[K - \underbrace{\frac{\mu}{\alpha}\left(1 - B(K-1, \frac{\mu}{\alpha})\right)}_{\text{mean \# in M/M/K-1/K-1}} \right] \\
&= \frac{K}{\mu} - \frac{1}{\alpha}\left(1 - B\left(K-1, \frac{\mu}{\alpha}\right)\right)
\end{aligned}$$

From Problem 9.23

$$B\left(K-1, \frac{\mu}{\alpha}\right) = \frac{\frac{\alpha K}{\mu} B\left(K, \frac{\mu}{\alpha}\right)}{1 - B\left(K, \frac{\mu}{\alpha}\right)}$$

$$\mathcal{E}[T] = \frac{K}{\mu} - \frac{1}{\alpha} + \frac{\frac{K}{\mu}B\left(K,\frac{\mu}{\alpha}\right)}{1 - B\left(K,\frac{\mu}{\alpha}\right)}$$

$$\mathcal{E}[T] = \frac{K}{\mu}\left[1 + \frac{B\left(K,\frac{\mu}{\alpha}\right)}{1 - B\left(K,\frac{\mu}{\alpha}\right)}\right] - \frac{1}{\alpha}$$

$$= \frac{K}{\mu}\frac{1}{1 - B\left(K,\frac{\mu}{\alpha}\right)} - \frac{1}{\alpha}$$

But for Problem 9.29 solution

$$\rho = \frac{\lambda}{\mu} = 1 - B\left(K,\frac{\mu}{\alpha}\right)$$

$$\Rightarrow \mathcal{E}[T] = \frac{K}{\lambda} - \frac{1}{\alpha} \quad \text{as desired} \quad \checkmark$$

9.6 M/G/1 Queueing Systems

9.33 A k-Erlang RV X with parameter k and λ has

$$\mathcal{E}[X] = \frac{k}{\lambda} \quad VAR[X] = \frac{k}{\lambda^2}$$

Since $\mathcal{E}[X] = \frac{1}{\mu}$ we have that $\lambda = k\mu$ and

$$VAR[X] = \frac{k}{k^2\mu^2} = \frac{1}{k\mu^2}$$

$$\Rightarrow C_X^2 = \frac{VAR[X]}{\mathcal{E}[X]^2} = \frac{1}{k}$$

$$\therefore \mathcal{E}[W]_{M/E_k/1} = \frac{\rho(1 + C_X^2)}{2(1-\rho)}\mathcal{E}[\tau] = \frac{\rho\left(1 + \frac{1}{k}\right)}{2(1-\rho)}\mathcal{E}[\tau]$$

For M/M/1 we let $k = 1$ and obtain

$$\mathcal{E}[W]_{M/M/1} = \frac{2\rho}{2(1-\rho)}\mathcal{E}[\tau]$$

For M/D/1 $C_X^0 = 0$ so

$$\mathcal{E}[W]_{M/D/1} = \frac{\rho}{2(1-\rho)}\mathcal{E}[\tau]$$

$$\therefore \mathcal{E}[W]_{M/D/1} < \mathcal{E}[W]_{M/E_k/1} \leq \mathcal{E}[W]_{M/M/1}$$

Since $\mathcal{E}[T] = \mathcal{E}[W] + \mathcal{E}[\tau]$ the same ordering applies for total delay.

9.38 a) The total time required to service a job is:

$$\tau = X + \sum_{i=1}^{N(X)} R_i$$

where $N(X)$ is the total number of times the machine breaks down. To find $C[\tau]$ we use conditional expectation:

$$\mathcal{E}[\tau] = \mathcal{E}[\mathcal{E}[\tau|X]]$$

$$\mathcal{E}[\tau|X=t] = t + \mathcal{E}\left[\sum_{i=1}^{N(t)} R_i\right] = t + \alpha t \mathcal{E}[R] \quad \text{from Eq. 5.13}$$

$$\Rightarrow \mathcal{E}[\tau] = \mathcal{E}[X + \alpha X \mathcal{E}[R]] = \underbrace{\mathcal{E}[X] + \alpha \mathcal{E}[X]\mathcal{E}[R]}_{\mathcal{E}[X](1+\alpha\mathcal{E}[R])} = \frac{1}{\mu}\left[1 + \frac{\alpha}{\beta}\right]$$

We also use conditional expectation to find $E[\tau^2]$:

$$\mathcal{E}[\tau^2] = \mathcal{E}[\mathcal{E}[\tau^2|X]]$$

$$\mathcal{E}[\tau^2|X=t] = \mathcal{E}\left[\left(t + \sum_{i=1}^{N(t)} R_i\right)^2\right]$$

$$= t^2 + 2t\mathcal{E}\left[\sum_{i=1}^{N(t)} R_i\right] + \mathcal{E}\left[\left(\sum_{i=1}^{N(t)} R_i\right)^2\right]$$

$$= t^2 + 2t(\alpha t + \mathcal{E}[R]) + \mathcal{E}\left[\left(\sum_{i=1}^{N(t)} R_i\right)^2\right]$$

$$\mathcal{E}\left[\left(\sum_{i=1}^{N(t)} R_i\right)^2\right] = \mathcal{E}\left[\mathcal{E}\left[\left(\sum_{i=1}^{N(t)} R_i\right)^2 \Big| N(t)\right]\right]$$

$$\mathcal{E}\left[\left(\sum_{i=1}^{N(t)} R_i\right)^2 \Big| N(t) = k\right] = \mathcal{E}\left[\sum_{i=1}^{k}\sum_{j=1}^{k} R_i R_j\right]$$

$$= k\mathcal{E}[R^2] + (k^2 - k)\mathcal{E}[R]^2$$

9.6. M/G/1 Queueing Systems

$$\therefore \mathcal{E}\left[\left(\sum_{i=1}^{N(t)} R_i\right)^2\right] = \mathcal{E}[N(t)\mathcal{E}[R^2] + [N^2(t) - N(t)]\mathcal{E}[R]^2]$$

$$= \mathcal{E}[N(t)]\mathcal{E}[R^2] + (\mathcal{E}[N^2(t)] - \mathcal{E}[N(t)])\mathcal{E}[R]^2$$

$$= \alpha t \mathcal{E}[R^2] + (\alpha t + (\alpha t)^2 - \alpha t)\mathcal{E}[R]^2$$

$$= \alpha t \mathcal{E}[R^2] + \alpha^2 t^2 \mathcal{E}[R]^2$$

$$\therefore \mathcal{E}[\tau^2 | X = t] = t^2 + 2\alpha t^2 \mathcal{E}[R] + \alpha t \mathcal{E}[R^2] + \alpha^2 t^2 \mathcal{E}[R]^2$$

finally

$$\mathcal{E}[\tau^2] = \mathcal{E}[X^2 + 2\alpha X^2 \mathcal{E}[R] + \alpha X \mathcal{E}[R^2] + \alpha^2 X^2 \mathcal{E}[R]^2]$$

$$= \mathcal{E}[X^2] \underbrace{[1 + 2\alpha \mathcal{E}[R] + \alpha^2 \mathcal{E}[R]^2]}_{(1+\alpha\mathcal{E}[R])^2} + \mathcal{E}[X]\alpha \mathcal{E}[R^2]$$

$$VAR[\tau] = \mathcal{E}[\tau^2] - \mathcal{E}[\tau]^2$$

$$= \mathcal{E}[X^2](1 + \alpha \mathcal{E}[R])^2 + \mathcal{E}[X]\alpha \mathcal{E}[R^2] - \mathcal{E}[X]^2(1 + \alpha \mathcal{E}[R])^2$$

$$= VAR[X](1 + \alpha \mathcal{E}[R])^2 + \mathcal{E}[X]\alpha \mathcal{E}[R^2]$$

$$= \frac{1}{\mu^2}\left(1 + \frac{\alpha}{\beta}\right)^2 + \frac{\alpha}{\mu}\frac{2}{\beta^2}$$

b) The coefficient of variation of τ is:

$$C_\tau^2 = \frac{VAR[\tau]}{\mathcal{E}[\tau]^2} = \frac{\frac{1}{\mu^2}\left(1 + \frac{\alpha}{\beta}\right)^2 + \frac{\alpha}{\mu}\frac{2}{\beta^2}}{\frac{1}{\mu^2}\left(1 + \frac{\alpha}{\beta}\right)^2} = 1 + \frac{2\alpha}{(\alpha+\beta)^2}$$

Thus the mean delay in the system is

$$\mathcal{E}[T] = \mathcal{E}[\tau] + \mathcal{E}[\tau]\frac{\rho}{2(1-\rho)}(1 + C_\tau^2)$$

$$= \mathcal{E}[\tau]\left[1 + \frac{\rho}{(1-\rho)}\left(1 + \frac{\alpha}{(\alpha+\beta)^2}\right)\right]$$

where

$$\rho = \lambda \mathcal{E}[\tau] = \frac{\lambda}{\mu}\left[1 + \frac{\alpha}{\beta}\right]$$

9.39 a) The proportion of time that the server works on low priority jobs is

$$\rho'_2 = 1 - \rho_1 = \lambda'_2 \mathcal{E}[\tau_2]$$
$$\Rightarrow \lambda'_2 = \frac{1 - \rho_1}{\mathcal{E}[\tau_2]} = \frac{1 - \lambda_1 \mathcal{E}[\tau_1]}{\mathcal{E}[\tau_2]}$$

b) From (9.105)

$$\begin{aligned}
\mathcal{E}[W_1] &= \frac{\lambda_1 \mathcal{E}[\tau_1^2] + \lambda'_2 \mathcal{E}[\tau_2^2]}{2(1 - \rho_1)} \\
&= \frac{\lambda_1 \mathcal{E}[\tau_1^2]}{2(1 - \rho_1)} + \frac{\lambda'_2 \mathcal{E}[\tau_2^2]}{2(1 - \rho_1)} \quad \text{but } \rho_1 = \lambda_1 \mathcal{E}[\tau_1],\ 1 - \rho_1 = \lambda'_2 \mathcal{E}[\tau_2] \\
&= \frac{\frac{\lambda_1}{2} \mathcal{E}[\tau_1^2]}{1 - \lambda_1 \mathcal{E}[\tau_1]} + \frac{\mathcal{E}[\tau_2^2]}{2\mathcal{E}[\tau_2]}.
\end{aligned}$$

9.7 M/G/1 Analysis Using Embedded Markov Chain

9.45 $\rho = \frac{\lambda}{\mu} = \left(\frac{\mu}{2}\right)/\mu = \frac{1}{2}$

a) For an M/G/1 system we have:

$$G_N(z) = \frac{(1 - \rho)(z - 1)\hat{\tau}(\lambda(1 - z))}{z - \hat{\tau}(\lambda(1 - z))}$$

where

$$\hat{\tau}(\lambda(1 - z)) = \frac{4\mu^2}{(s + 2\mu)^2}\bigg|_{s = \lambda(1 - z)} = \frac{4\mu^2}{(\lambda - \lambda z + 2\mu)^2}$$

$$\begin{aligned}
\Rightarrow G_N(z) &= \frac{\left(1 - \frac{1}{2}\right)(z - 1)4\mu^2}{z(\lambda - \lambda z + 2\mu) - 4\mu^2} = \frac{8}{z^2 - 9z + 16} \\
&\quad \text{where we used the fact that } \frac{\lambda}{\mu} = \frac{1}{2} \\
&= \frac{8}{(z - z_1)(z - z_2)} \quad z_1 = \frac{9 + \sqrt{17}}{2} \quad z_2 = \frac{9 - \sqrt{17}}{2} \\
&= \frac{8/z_1 z_2}{\left(1 - \frac{z}{z_1}\right)\left(1 - \frac{z}{z_2}\right)} = \frac{\frac{1}{2}}{\left(1 - \frac{1}{z_1}z\right)\left(1 - \frac{1}{z_2}z\right)}
\end{aligned}$$

9.7. M/G/1 Analysis Using Embedded Markov Chain

$$= \frac{A}{1 - \frac{1}{z_1}z} + \frac{B}{1 - \frac{1}{z_2}z} \Rightarrow \begin{array}{l} A = \frac{-z_2/2}{z_1 - z_2} \\ B = \frac{z_1/2}{z_1 - z_2} \end{array} \quad \text{partial fraction expansion}$$

$$= \frac{z_1/2}{(z_1 - z_2)\left(1 - \frac{1}{z_2}z\right)} = \frac{z_2/2}{(z_1 - z_2)\left(1 - \frac{z}{z_1}\right)}$$

$$= \frac{1}{2(z_1 - z_2)}\left[z_1 \sum_{j=0}^{\infty} \left(\frac{z}{z_2}\right)^j - z_2 \sum_{j=0}^{\infty} \left(\frac{z}{z_1}\right)^j\right]$$

$$\therefore P[N = j] = \frac{z_1}{2(z_1 - z_2)}\left(\frac{1}{z_2}\right)^j - \frac{z_2}{2(z_1 - z_2)}\left(\frac{1}{z_1}\right)^j \quad \text{coefficient of } z^j$$

$$P[N = j] = \frac{9 + \sqrt{17}}{4\sqrt{17}}\left(\frac{2}{9 - \sqrt{17}}\right)^j - \frac{9 - \sqrt{17}}{4\sqrt{17}}\left(\frac{2}{9 + \sqrt{17}}\right)^j$$

$$= \frac{8}{\sqrt{17}}\left(\frac{2}{9 - \sqrt{17}}\right)^{j+1} - \frac{8}{\sqrt{17}}\left(\frac{2}{9 + \sqrt{17}}\right)^j \quad j = 0, 1, \ldots$$

b) The Laplace Transform of the waiting time is:

$$\hat{W}(s) = \frac{(1 - \rho)s}{s - \lambda + \lambda \hat{r}(s)} = \frac{\frac{1}{2}s}{s - \lambda + \frac{\lambda 4\mu^2}{(s+2\mu)^2}} = \frac{1}{2}\left[\frac{s^2 + 8\lambda s + 16\lambda^2}{s^2 + 7\lambda s + 8\lambda^2}\right]$$

$$= \frac{1}{2}\left[1 + \frac{\left(\frac{\sqrt{17}+9}{2\sqrt{17}}\right)\lambda}{s + \left(\frac{7-\sqrt{17}}{2}\right)\lambda} + \frac{\left(\frac{\sqrt{17}-9}{2\sqrt{17}}\right)\lambda}{s + \left(\frac{7+\sqrt{17}}{2}\right)\lambda}\right]$$

$$f_W(t) = \mathcal{L}^{-1}[\hat{W}(s)]$$

$$= \frac{1}{2}\delta(t) + \frac{1}{2}\left(\frac{\sqrt{17}+9}{2\sqrt{17}}\right)\lambda e^{-\left(\frac{7-\sqrt{17}}{2}\right)\lambda t}u(t)$$

$$+ \frac{1}{2}\left(\frac{\sqrt{17}-9}{2\sqrt{17}}\right)\lambda e^{-\left(\frac{7+\sqrt{17}}{2}\right)\lambda t}u(t)$$

The total delay transform is:

$$\hat{T}(s) = \frac{(1 - \rho)s\hat{r}(s)}{s - \lambda + \lambda \hat{r}(s)} = \frac{\frac{1}{2}s\frac{4\mu^2}{(s+2\mu)^2}}{s - \lambda + \lambda\frac{4\mu^2}{(s+2\mu)^2}}$$

$$= \frac{8\lambda^2}{s^2 + 7\lambda s + 8\lambda^2}$$

$$\hat{T}(s) = \frac{8\lambda}{\sqrt{17}}\left[\frac{1}{s + \left(\frac{7-\sqrt{17}}{2}\right)\lambda} - \frac{1}{s + \left(\frac{7+\sqrt{17}}{2}\right)\lambda}\right]$$

$$f_T(t) = \mathcal{L}^{-1}[\hat{T}(s)] = \frac{8\lambda}{\sqrt{17}}\left[e^{-\left(\frac{7-\sqrt{17}}{2}\right)\lambda t} - e^{-\left(\frac{7+\sqrt{17}}{2}\right)\lambda t}\right]u(t)$$

9.8 Burke's Theorem: Departures from M/M/c Systems

9.51 a) If a departure leaves the system nonempty, then another customer commences service immediately. Thus the time until the next departure is an exponential random variable with mean $1/\mu$.

b) If a departure leaves the system empty, then the time until the next departure is equal to the sum of an exponential interarrival time (of mean $1/\lambda$) followed by an exponential service time (of mean $1/\mu$).

c) The Laplace transform of the interdeparture time is

$\frac{\mu}{s+\mu}$ when a departure leaves system nonempty

$\frac{\lambda}{s+\lambda}\frac{\mu}{s+\mu}$ when a departure leaves system empty

$$\therefore \mathcal{E}[e^{-sT_d}] = \frac{\mu}{s+\mu}\underbrace{\rho}_{\substack{\text{prob. system}\\\text{left nonempty}}} + \frac{\lambda}{s+\lambda}\frac{\mu}{s+\mu}\underbrace{(1-\rho)}_{\substack{\text{prob. system}\\\text{left empty}}}$$

$$= \frac{\lambda}{s+\mu} + \frac{\lambda(\mu-\lambda)}{(s+\lambda)(s+\mu)} = \frac{\lambda(s+\lambda)+\lambda\mu-\lambda^2}{(s+\lambda)(s+\mu)}$$

$$= \frac{\lambda}{s+\lambda} \Rightarrow T_d \text{ exponential with mean } 1/\lambda$$

9.9 Networks of Queues: Jackson's Theorem

9.56 a) $I = 3$

$$\left.\begin{array}{l}\pi_0 = p\pi_0 + \pi_1 + \pi_2\\ \pi_1 = \frac{1}{2}(1-p)\pi_0\\ \pi_2 = \frac{1}{2}(1-p)\pi_0\end{array}\right\} \quad \begin{array}{l}\pi_0 = \frac{1}{2-p}\\ \pi_1 = \pi_2 = \frac{1-p}{2(2-p)}\end{array}$$

9.9. Networks of Queues: Jackson's Theorem

Then

$$\lambda_0 = \lambda(3)\pi_0 = \frac{\lambda(3)}{2-p} \qquad \rho_0 = \frac{\lambda_0}{\mu}$$

$$\lambda_1 = \lambda(3)\pi_1 = \frac{\lambda(3)(1-p)}{2(2-p)} \qquad \rho_1 = \frac{\lambda_1}{\mu_1}$$

$$\lambda_2 = \lambda_1 \qquad \rho_2 = \frac{\lambda_1}{\mu_2}$$

$$\begin{aligned}
S(3) &= (1-\rho_0)(1-\rho_1)(1-\rho_2)[\rho_0^3 + \rho_1^3 + \rho_2^3 + \rho_0\rho_1^2 + \rho_0\rho_2^2 \\
&\quad + \rho_1\rho_2^2 + \rho_1\rho_0^2 + \rho_2\rho_0^2 + \rho_2\rho_1^2 + \rho_0\rho_1\rho_2] \\
&= (1-\rho_0)(1-\rho_1)(1-\rho_2)[(\rho_0^2 + \rho_1^2 + \rho_2^2)(\rho_0 + \rho_1 + \rho_2) + \rho_0\rho_1\rho_2]
\end{aligned}$$

$$\therefore P[N_0 = i, N_1 = j, N_2 = 3-i-j] = \frac{\rho_0^i \rho_1^j \rho_2^{3-i-j}}{(\rho_0^2 + \rho_1^2 + \rho_2^2)(\rho_0 + \rho_1 + \rho_2) + \rho_1\rho_2\rho_3}$$

$$0 \le i, j \quad \text{and} \quad i+j \le 3$$

b) The program completion rate is

$$p\mu[1 - P[N_0 = 0]] = p\mu \frac{\rho_0^3 + \rho_0^2\rho_1 + \rho_0^2\rho_2 + \rho_0\rho_2^2 + \rho_0\rho_1^2 + \rho_0\rho_1\rho_2}{(\rho_0^2 + \rho_1^2 + \rho_2^2)(\rho_0 + \rho_1 + \rho_2) + \rho_1\rho_2\rho_3}$$